I0063620

THE COMING ECONOMIC REVOLUTION

Written by
Donald D. Dienst

DienstNet
Where imagination meets reality

Dear Reader,

Thank you for choosing this book. If you enjoyed your reading journey, please consider leaving a review. Your thoughts and opinions are invaluable, not just to us but to fellow readers as well. Reviews help guide others to books they will love and cherish. Your feedback is a beacon in the vast sea of literature, guiding like-minded readers to their next great read. Here are a few places where your review can make a significant impact:

> **Amazon:** A key destination for book lovers, your review here can reach a wide and diverse audience.

> **Goodreads:** As a vibrant community of readers, your thoughts can spark discussions and influence book choices.

> **Barnes & Noble:** Your review on this major book-selling platform can guide fellow readers in their selections.

Every review helps in building a community of passionate readers and in sharing the joy of reading. Thank you for your time and thoughts.

Happy Reading!

THE COMING ECONOMIC REVOLUTION

Written by
Donald D. Dienst

DienstNet
Where imagination meets reality

Special Offer for Physical Book Owners!

Register your physical book purchase at www.DienstNet.com to gain exclusive access to the audio version of this book, free of charge. Simply upload a photo of this copyright page with your first and last name clearly printed on it to verify your purchase. Enhance your learning and understanding, on us!

The Coming Economic Revolution

Copyright 2023 by DienstNet LLC
All Rights Reserved.
No part of this publication may be reproduced, distributed, or transmitted in any form or by any means, including photocopying, recording, or other electronic or mechanical methods, without the prior written permission of the author, except in the case of brief quotations embodied in critical reviews and certain other noncommercial uses permitted by copyright law.

SOFTCOVER
FIRST EDITION
ISBN: 979-8-9895994-0-0
LCCN: 2023951642

Information has been obtained by DienstNet LLC from sources believed to be reliable. However, because of the possibility of human or mechanical error by our sources, DienstNet LLC, does not guarantee the accuracy, adequacy, or completeness of any information and is not responsible for any errors or omissions or the results obtained from the use of such information. All brand names and product names mentioned in this book are trademarks or service marks of their respective companies. Any omission or misuse (of any kind) of service marks or trademarks should not be regarded as intent to infringe on the property of others. The publisher recognizes and respects all marks used by companies, manufacturers, and developers as a means to distinguish their products.

Published by DienstNet LLC
3 The Trail North
Hackettstown, NJ 07840
info@DienstNet.com
www.DienstNet.com

TABLE OF CONTENTS

Section I The Foundation

Section II Emergence of a New Economic Actor

Section III Understanding Bitcoin

Section IV Emerging Applications and Challenges

Section V Looking Ahead

Appendix

Section I
The Foundation

Chapter 1
Introduction

Money is eternal, but it doesn't last forever. This enigmatic paradox lies at the heart of our economic narrative, where the concept of value exchange has remained a constant since the earliest civilizations. Yet, the physical embodiments of currency are marked by their inevitable demise. The dollar, the cornerstone of contemporary finance, teeters on the edge of this historical inevitability. Here, we stand at the cusp of a decisive moment as the twilight of an old monetary epoch approaches and the murmurs of a new economic dawn become unmistakable. Amid this transformative era, Bitcoin emerges as a contender, poised to redefine what we accept as 'money.'

The enduring presence of money belies a growing unrest beneath the surface of our global economic landscape—a widespread dissatisfaction with prevailing economic models. Once reliable maps of fiscal reality, these models now seem ill-equipped to chart the treacherous terrains of today's financial ecosystems. The inadequacies of these outdated paradigms are not merely academic concerns; they profoundly influence billions of lives, shaping the policies that govern markets, nations' prosperity, and societies' well-being. Amidst such sweeping influence, the call for a paradigmatic shift grows urgent. It is no longer a question of if but when and how we will undertake the formidable task of re-evaluating and reconstructing the economic principles that underpin our world—a task that may dictate the trajectory of our collective future.

As we stand at the crossroads of financial tradition and innovation, a critical interrogation of economic orthodoxy becomes relevant and imperative. The pages ahead promise a deep dive into the very bedrock of conventional economic thought, challenging doctrines that have long gone unchallenged. What are the hidden flaws within the theories we've so staunchly relied upon? How have they shaped our current financial realities, for better or worse? And most importantly, what alternatives lie beyond the horizons of traditional economics? These questions serve as our compass, guiding us through a journey of intellectual discovery and critical analysis, beckoning us to unravel the complex tapestry of modern economics.

Within this book, we navigate a carefully structured exploration, laying out a roadmap that begins in the past and forges into the future. Each chapter unfolds sequentially, building upon the last to establish a comprehensive narrative. We start by dissecting the failings of traditional economic theories, setting the historical context from which modern economic thought has emerged. As we progress, the focus shifts from the evolution of money to the revolutionary inception of Bitcoin, marking the transition from historical examination to contemporary relevance.

Further chapters delve into Bitcoin's role as digital gold, its technical underpinnings, and its potential as a new medium of exchange, steadily drawing the line from theory to practical application. The discussion of decentralization and trustlessness is real-world evidence of the technology's transformative power. As criticisms and challenges are laid bare, the robustness of the argument for change is tested and strengthened.

The final chapters cast a vision of the future, considering Bitcoin's place within an emerging economic paradigm. The structured progression of the book actively ensures that readers, by its conclusion, are informed about the historical and theoretical facets of our current economic moment and equipped to understand and engage with the call for a sweeping financial overhaul. Each chapter is a puzzle piece, and when connected, they reveal the compelling picture of why and how a revolution in economic thought is upon us.

In this treatise, Bitcoin is presented not merely as a novel form of digital currency but as a symbol of the seismic shift possible within economics.

Bitcoin serves as a harbinger of change, standing as a testament to the potential for decentralization, transparency, and resilience in a system traditionally governed by centralization and opacity. The narrative woven through the chapters carefully positions Bitcoin as more than a technological marvel; it is a catalyst for reimagining the very foundations of economic interaction and value exchange.

As we journey through the discourse, Bitcoin continually assumes a transformative role, consistently portrayed in this light. It prompts us to reconsider fundamental economic questions: What should money look like in a digital age? Can we achieve a global economy free from central authorities' whims and accessible to all? By anticipating the convergence of economic theory and this emergent technology, the book foreshadows a future where Bitcoin could be at the heart of a radically different economic landscape, shaping how we understand, utilize, and value the concept of currency.

This discussion transcends academic theory and enters the realm of daily life, where the implications of our economic system touch every purchase, every job, every savings account, and the broader future we are crafting. The chapters within this book are not merely a reflection on an abstract economic revolution; they are a guide to understanding how the impending changes will affect your financial autonomy, the stability of your currency, and the global economic landscape that influences your everyday decisions.

Presenting this book as a navigational tool, it aims to equip you, the reader, with the knowledge to understand the shifts underfoot and the insight to adapt to them. This discourse seeks to empower you with insights that enable informed decisions, whether you are investing in cryptocurrency or simply curious about the future of money. The conversation within these pages is about laying a foundation for readers to anticipate and actively participate in the changes to come, ensuring that you are prepared to move forward in a world where the economic ground is shifting beneath us all.

As each page turns, you receive a cordial invitation to embark on a voyage of economic discovery, equipped with an open mind, ready to question, and eager to understand. This book is not just a compilation of chapters; it is an odyssey through the intricacies of our current financial system and a look into the possibilities of tomorrow. Let's challenge the precepts of the past and allow the ideas of the future to inspire as we move from the roots of traditional economic thought to the burgeoning branch of Bitcoin and its kin.

Prepare to engage with complex concepts and visionary ideas that will provoke thought and, possibly, debate. Consider this a call to intellectual arms, to arm yourselves with knowledge and insight that could redefine your interaction with the world's economic gears. This journey is as much yours as the global economy's, for each individual plays a role in the market's tides. As you turn the pages, ponder your place in this sweeping narrative. What will your role be in the impending economic revolution? How will you adapt, influence, and prosper within this new era? The chapters ahead are more than a story; they are an invitation to join the revolution.

Chapter 2
The Evolution of Money

In the story of civilization, money has played the starring role. It is more than mere currency—it is a medium that facilitates exchange, a measure that assigns value, and a means to store wealth. From the earliest days of human society, the need for a medium of exchange was clear; bartering, the direct trade of goods and services, was limited by the need for a 'coincidence of wants.' Money emerged as the ingenious solution, evolving from primitive forms to today's complex digital currencies, fundamentally changing how we interact and transact.

The journey of money began with bartering, where goods were traded directly—a system fraught with limitations, not least the difficulty of finding mutually desirable exchanges. The quest for a more flexible medium led to the adoption of various commodities as money. Livestock and agricultural produce served this purpose for a time but were soon supplanted by something more enduring: metal. Precious metals, owing to their rarity and durability, became the standard. They were shaped into coins, each stamped with a mark to verify its weight and substance. They gave rise to the first currencies and altered trade and economies irreversibly.

Paper money marked the next great leap. Born of necessity in the Tang Dynasty, when carrying heavy coinage became impractical, these paper notes promised redemption in metal but soon gained acceptance as money in their own right. As trade routes expanded, so did the concept of paper money, spreading to every corner of the globe.

Yet, the connection to physical metal was not to last. Fiat money, a currency without intrinsic value, decreed legal tender by governments, became the norm. This transformation was complete with the abandonment of the gold standard; money's value was now a matter of national decree, not metallic content. Central banks emerged as the shepherds of these fiat currencies, tasked with balancing the money supply against the economy's health.

The story of fiat money is not without its quirks. Its value, unanchored to any physical commodity, is vulnerable to inflation, where money's purchasing power erodes over time, a specter that haunts economies to this day. Governments with the power to print money can fall to the temptation of manipulating the money supply for political ends, sometimes disastrously so. History is punctuated with episodes of hyperinflation and currency collapse, stark reminders of fiat money's fragility.

Against this backdrop rises the concept of 'sound money'—money that is stable, retains its value over time, and is difficult to manipulate. Historically, sound money has been synonymous with economic stability, its discipline a bulwark against the vicissitudes of inflation and political meddling. In today's climate of financial uncertainty, the call for a return to sound money principles grows ever louder. It is within this crucible of fiscal discontent that alternatives like Bitcoin emerge, promising a return to many of the virtues of sound money through their inherent design.

Chapter 3
The Cracks in the Foundation

From Mercantilism to Corporate Capitalism

Mercantilism, which dominated European economic policy from the 16th to the 18th centuries, was not just an economic theory but also a practice that underpinned the very fabric of society during its time. Its essence was the belief that the wealth of nations was based on the amount of precious metals—gold and silver—they accumulated. Under this system, a country's economic health was measured by its stockpile of these metals, considered the ultimate source of economic power.

Countries adhered to a policy of accumulating wealth through trade surplus, encouraging exports, and discouraging imports through tariffs and monopolies. Colonies became crucial in this framework, serving as both the sources of raw materials and the markets for manufactured goods from the mother country. The flow of wealth from colony to colonizer was seen as a zero-sum game, where one nation's gain was another's loss.

The historical context of mercantilism coincided with the Age of Exploration, where European powers expanded their global reach. Ships returned from distant lands filled with precious metals and exotic commodities, fueling the mercantilist thirst for accumulation. This era saw the rise of national treasuries and the beginnings of what would eventually evolve into the modern banking system.

But as much as mercantilism spurred global exploration and trade, it also planted the seeds of its demise. The relentless pursuit of a positive trade balance frequently led to conflicts as nations vied for control over resources and markets. Moreover, the heavy regulation and control over economies stifled innovation and disregarded the consumer's welfare for the state's wealth.

While mercantilism laid the groundwork for a global economy, the limitations of this system—its propensity for conflict, its neglect of individual enterprise, and the finite nature of precious metal accumulation—set the stage for the development of new economic theories. These theories would eventually recognize the market forces and the potential for wealth creation through industrial innovation, setting the course toward capitalism and beyond.

The shadow of mercantilism, however, did not completely fade. Echoes of its principles can be seen in modern economic policies prioritizing trade surpluses and state intervention. Yet, the shift from mercantilism's focus on state accumulation of wealth to the individual's role in wealth generation marks the actual turning point, paving the way for Adam Smith's revolutionary ideas on capitalism, which would redefine the economic landscape of the modern world.

Adam Smith's Vision of Capitalism

Adam Smith's treatise, "The Wealth of Nations," published in 1776, marked a paradigmatic shift from the mercantile practices of the time to a new vision that would eventually be known as capitalism. Smith introduced a model where the self-regulating behavior of the marketplace, guided by individuals' pursuit of their own interests, would lead to economic prosperity and societal benefit—a concept famously referred to as "the invisible hand."

Smith's model of capitalism was rooted in the belief that economic success was not derived from the accumulation of commodities, like gold and silver, but from the production of goods and services and their exchange in a free market. He championed laissez-faire economics, a system with minimal government intervention, where the natural dynamics of supply and demand determined the price and distribution of goods.

The beauty of Smith's "invisible hand" was that it turned personal self-interest into a collective good. As each individual strove to maximize their own gain, they inadvertently contributed to the overall economic welfare of society. This resulted in a more efficient allocation of resources, as producers and consumers made rational decisions based on their own interests.

However, even with its significant benefits, Smith's capitalist model was not without its flaws. Critics argue that unchecked self-interest can lead to monopolies and unethical business practices that harm society. There is also the issue of externalities—costs or benefits of economic activities not reflected in market prices—such as pollution, which the free market does not address on its own. Moreover, the model's assumption that all market participants act rationally and possess equal information has been contested.

As the 19th century unfolded, Smith's theoretical model of capitalism was tested during the Industrial Revolution. This period saw unprecedented economic growth, technological advancement, and significant social upheaval. The harsh realities of industrial life exposed the gaps in Smith's theory, particularly regarding the welfare of the working class and the distribution of wealth.

The era highlighted the need to balance the free market's efficiencies and regulatory measures to protect laborers and ensure fair competition. It was a time of economic and social experimentation, leading to the development of new theories and policies that sought to address the deficiencies of pure capitalist doctrine and, ultimately, to the birth of corporate capitalism, which promised growth and efficiency on an even larger scale.

The Rise of Corporatism

Corporatism represents a significant divergence from the capitalism envisioned by Adam Smith. While Smith's model emphasized the roles of individual entrepreneurs and small businesses as the drivers of economic growth, corporatism shifted the focus toward large corporations as the central actors. These entities, legally defined as individuals but possessing far greater resources and influence than anyone, became the economic landscape's new titans.

Before New Jersey's pivotal legal changes in the late 19th century, corporations in the United States were chartered by state governments for specific purposes and had limited lifespans. These entities were seen as extensions of the public will, designed to fulfill public works projects or specific commercial ventures that were in the public interest. Corporations had their powers and operations tightly controlled by the state legislatures that granted their charters, with strict regulations on what they could and couldn't do.

However, this changed dramatically with the passage of New Jersey's legislature on corporate charters. In 1896, under the governorship of John W. Griggs, New Jersey passed laws that allowed companies to incorporate more easily and with fewer restrictions. Corporations could now be chartered in New Jersey without specific state legislative approval, and they were permitted to have perpetual duration, engage in any lawful business, and hold stock in other corporations.

These legislative amendments made New Jersey the first state to allow such freedoms for corporations. This move ushered in a new era of corporate governance, paving the way for the rise of corporate power as we understand it today.

The new laws in New Jersey appealed to business owners. They soon led to a wave of corporations being chartered in the state, thus becoming the precursor to a "holding company" and the genesis of corporate conglomerates. The ability for a corporation to own other corporations without legislative approval opened the door for the creation of trusts and monopolies, which would dominate many sectors of the American economy and exert considerable influence over national and international commerce.

This transformation marked the shift from Adam Smith's vision of capitalism, based on competition and free markets, to a landscape dominated by large corporate entities with significant economic and political power. This profound change in the corporate landscape, stemming from the laws passed in New Jersey, became a critical milestone in the evolution of American and, eventually, global economic policy.

The period after 1896 heralded the dawn of corporatism, mainly in the United States, with the ascent of trusts and monopolies. These corporations amassed unprecedented capital, allowing them to swallow competitors, control vast market shares, and influence political processes. Unlike the local businesses of Smith's capitalism, these corporations operated nationally and eventually internationally, often prioritizing shareholder interests above those of workers, consumers, or the environment.

As they grew in power, corporations began to wield significant influence over economic policies. The legal concept that "money is speech," particularly in the U.S., allowed corporations to funnel vast sums into the political system, effectively buying influence and shaping legislation to suit their interests. This, in turn, paved the way for a 'revolving door' phenomenon, where individuals move between roles as legislators or regulators and positions within the corporations they once oversaw, creating conflicts of interest and policy biases.

Corporations are essentially 'eternal virtual beings,' outliving their founders and not bound by the natural lifespan of human beings. Their sole legally defined purpose, the enrichment of shareholders, often overshadows broader social and economic responsibilities, raising ethical questions about the role and power of such entities in society.

The infusion of corporate money into politics has led to a distortion of the democratic process, where the voices of ordinary citizens are at risk of being drowned out by the financial clout of corporate entities. This has also led to regulatory capture, where public agencies that are supposed to regulate industries end up being influenced by the very corporations they're tasked with overseeing.

The impact of corporatism is not confined to national borders. Multinational corporations have become critical players in the global economy, influencing trade agreements, labor standards, and even the economic policies of sovereign states. The spread of corporatism has thus raised concerns about economic homogenization, loss of cultural diversity, and the power of corporations to override local economic priorities.

Keynesian Economics and Its Global Influence

Keynesian economics represents a paradigm shift introduced by John Maynard Keynes, suggesting that active government intervention is crucial during economic downturns. Contrary to classical economics, which promoted minimal interference, Keynes argued that governments should counteract reduced private spending with public spending during recessions. This could involve increased government expenditures and tax cuts to stimulate demand, job creation, and economic stability.

The implementation of Keynesian economic policies significantly altered the landscape of government spending. With an emphasis on the government's role in moderating economic cycles, the restraint traditionally imposed on fiscal policy was effectively unlocked. This newfound flexibility in managing the economy also contributed to unraveling the gold standard. Keynes was critical of the gold standard, seeing it as a constraint on the economy that prevented governments from increasing the money supply during downturns. The Bretton Woods system, influenced by Keynesian thought, initially tied currencies to the U.S. dollar, which was convertible to gold.

However, the pressures of maintaining this system alongside expansive fiscal policies became unsustainable, leading to President Nixon's decision to end the dollar's convertibility into gold in 1971, a move that fully detached currencies from the gold standard and ushered in an era of fiat money.

This detachment from the gold standard opened the floodgates for an expansion in the money supply, as the amount of gold reserves no longer constrained governments. The ability to print money at will has since been a double-edged sword. At the same time, it allowed for a more responsive monetary policy. Still, it also led to concerns over currency debasement. As more money entered circulation without the backing of gold, the purchasing power of currencies began to erode, a trend that has been particularly pronounced in recent decades. Critics of Keynesian economics argue that its policies have facilitated this debasement, as the temptation to fund government spending through money creation rather than taxation or borrowing can be hard to resist.

The often-quoted Keynesian aphorism, "In the long run, we are all dead," encapsulates a certain pragmatism that underscored Keynes's approach to economic policy. This statement, which originated from his 1923 work "A Tract on Monetary Reform," was a critique of the classical economic belief in the self-correcting nature of markets, arguing that economic problems should be dealt with in the present rather than waiting for long-term market solutions which could lead to prolonged suffering.

While this perspective brought forward the concept of immediate government intervention to mitigate economic downturns, it also paved the way for criticisms of potential shortsightedness within Keynesian policy prescriptions.

Detractors argue that such an approach might disregard the long-term consequences of economic actions, particularly the impact on currency value. The disconnection from the gold standard, influenced by Keynesian thinking, removed a traditional restraint on expanding the money supply. The lack of tangible asset backing meant governments had greater liberty to print money, leading to fears of currency debasement. Critics suggest that the Keynesian focus on short-term economic relief could lead to long-term inflationary pressures, eroding the purchasing power of money and savings over time, a trend that has materialized in various economies over the past decades.

The brevity of Keynes's quip, in the long run, does not entirely encapsulate his views on the importance of future consequences; however, it has come to symbolize the critiques of Keynesian economics' perceived focus on short-term solutions. In the context of currency debasement, the quote has been used to highlight a tension within Keynesianism: the balancing act between addressing present economic woes and maintaining the integrity of a currency over time. The Keynesian willingness to engage in deficit spending and monetary expansion, especially during recessions, can be contrasted with the potential ramifications of these policies, including the debasement of currency if left unchecked.

The critique of Keynesianism grew stronger in the late 20th century, especially during periods of stagflation, which saw the combination of high inflation and unemployment. The Keynesian prescription of pumping government money into the economy was seen as contributing to persistent inflation and undermining the value of money. Moreover, the long-term impacts of sustained government deficits began to manifest, raising questions about fiscal sustainability and the national debt burden.

The transition away from gold-backed currency to a reliance on fiat money has led many to question the very foundation of modern monetary policy. This reflection on the consequences of Keynesian practices sets the stage for a discussion on returning to a more sustainable economic model, including alternatives like Bitcoin, which promises a fixed supply and a departure from the government-controlled monetary systems.

As we witness fluctuating economic cycles and debate the appropriate levels of government intervention, Keynes's words remind us of the delicate trade-off between short-term recovery tactics and their long-term sustainability. This discourse is integral to understanding the broader context of our current monetary system and the appeal of alternative economic models, such as that proposed by Bitcoin, which seeks to address such long-term concerns by offering a deflationary counterpoint to the fiat currency's inflationary tendencies.

Chapter 4
The End of a Currency

Currencies, while symbols of economic stability, are not impervious to demise. Their end can come about in several forms, each emblematic of the volatility that underlies nations' financial systems. Perhaps the most dramatic ending for a currency is through hyperinflation — a scenario where prices rise uncontrollably and the currency's value plummets to near worthlessness. Historical specters of hyperinflation, such as those haunting the Weimar Republic or, more recently, Zimbabwe, provide stark lessons on the fragility of fiat money. In such cases, currencies often become so devalued that they must be entirely abandoned or replaced through monetary reform.

Political upheavals can also spell the end for a currency. The collapse of a government, or the dissolution of an entire country, as witnessed with the fall of the Soviet Union, necessitates the creation of new currencies to cement economic sovereignty in emerging nation-states. Additionally, forming economic unions, such as the European Union, can merge monetary identities into a single currency, as with the euro supplanting the individual currencies of its member states.

Sometimes, the end of a currency is engineered through deliberate policy. Nations may peg their currency to a more stable one to import monetary discipline. However, should the economic winds prove too turbulent, such pegs can falter, triggering a currency crisis that demands a new monetary system. Governments, by decree or through sweeping legislation, may also retire a currency to introduce a more robust one or as part of a larger scheme of economic reforms.

Lastly, in the wake of natural disasters or wars, a currency may cease to function as a medium of exchange due to economic disruption, necessitating a new form of currency.

Therefore, the ending of a currency is often both a reflection and a determinant of a country's economic resilience and flexibility. Each instance provides a vivid narrative of economic adaptation, sometimes fraught with hardship but always indicative of change and the search for stability. When natural disasters or wars wreak havoc on a nation, the chaos can damage the established economic order, often leading to a currency's collapse. While a direct result of a catastrophe, this breakdown is exacerbated by governmental inadequacies. When a currency falls, it often reveals underlying weaknesses in governance — the lack of contingency planning, inadequate economic buffers, and poor disaster response mechanisms. Each fallen currency tells a tale of economic readjustment and governmental shortfall, marking a desperate scramble for monetary stability amidst the ruins left by government shortcomings.

The Perils of Currency Debasement

Currency debasement is as old as currency itself, with a legacy of devastating impacts on economies and the fabric of societies. At its core, debasement refers to the deliberate reduction in the value of a currency. In a historical context, this often involved decreasing the precious metal content in coins. In contemporary times, debasement is less about the alteration of physical money and more about the erosion of purchasing power through inflation, often due to excessive money printing by governments.

The annals of history are replete with instances of currency debasement. Ancient Rome is perhaps the most notable, where successive emperors chipped away at the silver content of their denarii to fund expansive—and expensive— imperial ambitions. This short-term solution led to long-term economic crises, including rampant inflation and a public loss of confidence in the currency's value. Across the channel, during the 16th century, England witnessed "The Great Debasement" under Henry VIII, whose significant reduction in the precious metal content of coins was an effort to cover his lavish expenditures and sustained military campaigns, ultimately leading to inflation and monetary instability.

In modern economies, debasement typically manifests through various inflationary mechanisms. Quantitative easing, a tool central banks use to stimulate the economy, can lead to debasement if the new money entering the system dilutes the currency's value in circulation. Similarly, deficit spending without a corresponding increase in production leads governments to print more money, thereby decreasing its overall value. Persistently low interest rates can also contribute to an expanded money supply, as cheap borrowing fuels consumer spending and credit expansion beyond sustainable economic growth.

Even the abandonment of traditional standards, such as the shift from the gold standard to fiat currency, has played a role. Without the gold standard's natural restriction on the money supply, governments gain far greater latitude to increase the quantity of money, with potentially disastrous inflationary consequences.

The repercussions of currency debasement are wide-ranging. Inflation, a typical result of debasement, acts as a silent thief, chipping away at the value of saved money and thereby harming retirees and those on fixed incomes. It deepens the chasm of income inequality, as those with assets that appreciate can hedge against inflation. At the same time, those with cash savings watch their purchasing power dwindle.

Moreover, debasement can sow the seeds of broader economic instability. During the hyperinflationary episodes of the Weimar Republic, Zimbabwe, and, more recently, Venezuela, the rapid loss of currency value led not just to economic turmoil but to social unrest and a loss of faith in the government's fiscal stewardship. These conditions discourage foreign investment as international investors seek economies with more stable monetary policies.

The modern era has seen its share of economic crises, many of which have been exacerbated, if not directly caused, by currency debasement. While multifaceted in its origins, the Global Financial Crisis of 2007-2008 was undoubtedly magnified by a period of loose monetary policy and easy credit. This trend of expansive monetary supply has, in some emerging markets, led to severe currency devaluation, precipitating capital flight and economic downturns. Even the most robust economies are not immune, as seen in instances where investor confidence falters in the face of aggressive monetary expansion.

Debasement has also contributed to a growing interest in alternative stores of value. To escape the vulnerabilities of fiat currencies, some investors and consumers have turned to commodities, real estate, and more recently, cryptocurrencies like Bitcoin, which offer a fixed supply and are thus perceived as immune to government-induced inflation in the long term.

Fractional Reserve Banking and Financial Instability

In the labyrinthine world of modern finance, few concepts are as simultaneously foundational and contentious as fractional reserve banking. At its core, this system permits banks to hold only a fraction of their depositors' funds in reserve, using the remainder to extend loans and generate profits. It's a practice that rests on the delicate assumption that not all depositors will seek to withdraw their funds simultaneously, a gamble that balances on the fulcrum of public confidence.

The historical adoption of fractional reserve banking was a gradual evolution from the practices of goldsmiths and early bankers who realized that they could safely lend out a portion of the gold and coins stored in their vaults without depleting their reserves. This realization seeded the modern banking model, enabling the exponential growth of credit and, by extension, the economy's expansion. As this method took root and sprouted into the complex banking operations of today, it inadvertently wove a pattern of potential economic overreach into the fabric of currency stability.

The bank runs of the 1800s serve as a textbook example, where rampant speculation, economic panic, and a pervasive lack of regulatory oversight led to widespread runs on banks. Depositors, driven by fear, would rush to withdraw their funds, only to find insufficient physical currency to fulfill their demands. The ensuing chaos often led to the collapse of otherwise solvent banks, sending ripples through the economy that would precipitate further panic. The resulting crises led to the collapse of banks and the rapid devaluation and sometimes total collapse of the very currencies they dealt in as public confidence evaporated.

These episodes underscore a critical aspect of currency life cycles. Expanding the money supply through fractional reserve banking can lead to inflationary pressures, diminishing the currency's purchasing power and contributing to its eventual debasement. In extreme cases, such as when compounded by other economic woes or policy missteps, it can precipitate the kind of currency crises that consign monetary systems to the history books.

In contemporary discourse, the lessons of the past color the present debates. As central banks worldwide engage in complex monetary policies, often expanding the money supply to counteract economic downturns, the specter of currency debasement looms large. Economists and policymakers wrestle with the question of balancing the advantages of fractional reserve banking with the primal need to prevent the erosion of currency value that can lead to an economy's undoing. Conversely, proponents maintain that when coupled with appropriate regulation and insurance mechanisms like the Federal Deposit Insurance Corporation (FDIC) in the United States, fractional reserve banking is a dynamic engine for economic growth.

Therefore, the continued existence of a currency is not just a matter of economic output or trade balances; it is intrinsically linked to the trust placed in the currency's value. As history has shown, once that trust begins to wane under the weight of an overextended banking system and a ballooning money supply, the end of a currency may be in sight. Thus, the story of fractional reserve banking is woven tightly into the tale of how currencies live and die, reminding us that the stability of our money reflects the prudence with which we manage the wellspring of its very creation.

The Case for a New Economic Paradigm

The historical limitations of these theories are not merely academic footnotes but are etched deeply into the economic landscape—often in the form of crises that recur with disheartening predictability. The gold standard, once a bedrock of fiscal stability, gave way under the weight of modern economic demands. At the same time, Keynesianism, though revolutionary, proved not to be the panacea for all economic downturns, as the stagflation of the 1970s and the financial crisis of 2008 bear witness. Even Adam Smith's invisible hand, which posits a self-regulating market, has been forced open to reveal the disparities and inefficiencies it can inadvertently nurture.

An analysis of these crises and their root causes—a complex interplay of unfettered speculation, lax regulation, and the unintended consequences of well-meaning policy—lays bare the need for a new economic paradigm. This paradigm must be resilient yet flexible, able to harness the innovative energies of the market while safeguarding against its excesses. The question of sustainability looms large: how do we create an economy that serves the needs of the present without compromising the well-being of future generations?

In the quest for such a paradigm, the revolutionary potential of Bitcoin and blockchain technology emerges as a beacon of potential change. With its decentralized nature and finite supply, Bitcoin challenges traditional notions of what a currency can be and offers a radical alternative to government-issued money. Could this digital asset, which operates outside the traditional banking system and beyond the reach of conventional monetary policy, represent a key piece in the puzzle of our economic future?

This is more than a rhetorical question; it's an invitation to explore the possibilities inherent in this nascent technology as we stand on the brink of what could be the most significant economic revolution since the advent of paper currency. Bitcoin beckons with the promise of a new chapter in the annals of economic history. This chapter remains to be written by us all. In the following pages, we shall peel back the layers of this cryptographic innovation, scrutinizing its principles, performance, and promise. Prepare to delve into the heart of Bitcoin, for it may just herald the dawn of a new economic epoch.

Section II
Emergence of a New Economic Actor

Chapter 5
The Birth of Bitcoin

In the sweeping saga of monetary evolution, from shells to digital bits, Bitcoin emerges as a new chapter and a potential redefinition of the currency itself. To contextualize Bitcoin in the annals of financial history, one must appreciate the transformative shifts that have occurred as societies moved from tangible assets to abstract representations of value. Bitcoin stands on the precipice of this historical continuum, a digital heir to the monetary throne, challenging the very architecture of how value is created, stored, and transferred.

The genesis of Bitcoin is rooted in a vision as old as money itself — the quest for a currency that is free from the whims of sovereigns and the shackles of borders. In the face of a centralized banking system, which became synonymous with control, Bitcoin was conceived as a bastion of decentralization. Bitcoin's creation was a deliberate and thoughtful response to a history of money that, time and again, had been manipulated to serve the few at the expense of the many.

Bitcoin's architecture eschews the centralized paradigm, instead opting for a dispersed network of nodes, each participant upholding and verifying the ledger — a public testament to the currency's movements and ownership. This framework ensures that no single entity can hijack the system, making Bitcoin an embodiment of decentralized philosophy. The vision is not just to create a new form of money but to revolutionize the concept of currency by returning power to the hands of the individuals who use it.

Thus, in the grand chronicle of financial instruments, Bitcoin is a maverick, a decentralized currency that challenges centuries of centralized monetary control. It is both a nod to the past — respecting the integrity of sound, tamper-proof money — and a bold leap into the future, heralding an era of financial autonomy unmediated by traditional gatekeepers.

Before the advent of Bitcoin, the digital currency landscape was a frontier of numerous but ultimately unsuccessful ventures. Experimenters and cryptographers had long been fascinated with creating a form of money native to the burgeoning digital realm. Various iterations of digital cash, such as DigiCash, Bit Gold, and b-money, attempted to harness the power of cryptography to secure transactions. However, these prototypes either failed to gain traction or fell victim to inherent flaws in their design, particularly their vulnerability to double-spending and reliance on some form of centralized authority.

Against this backdrop of trial and error, Bitcoin arrived as a solution to a long-standing problem. The publication of the Bitcoin whitepaper in 2008 was a watershed moment. It outlined a decentralized network that used a public ledger called a blockchain to record transactions. This system solved the double-spending problem through an innovative consensus mechanism known as proof-of-work. The simplicity and elegance of the paper suggested a profound understanding of the financial and cryptographic principles necessary to forge digital money.

The author of the whitepaper, known only as Satoshi Nakamoto, remained as enigmatic as the solution he, she, or they proposed was revolutionary. To this day, the true identity of Satoshi remains shrouded in mystery, a riddle wrapped in the enigma of cypherspace. Speculation about Nakamoto's identity has become a subculture within the Bitcoin community. Still, the creator's anonymity also stands as a philosophical statement — Bitcoin is not about the figure of a founder but about the empowerment of the collective.

The introduction of the Bitcoin whitepaper did not merely signal the birth of a new currency; it heralded a challenge to the established financial order. It was an act of technological rebellion, written not on paper but on the very fabric of the internet. In this document, the seeds of an economic revolution were sown, and from these seeds grew a mighty tree whose branches continue to spread, offering shelter and strength to those who understand its roots.

The Principles of Bitcoin

Within the digital heart of Bitcoin beats the pulse of several core principles, chief among them being decentralization. This term resonates with the promise of a financial system liberated from central authorities. Decentralization in the context of Bitcoin means that no single entity, be it a government, corporation, or individual, holds power over the network. This diffusion of control is pivotal, as it democratizes financial transactions, allowing individuals to interact directly with one another. The power, quite literally, rests with the people. Bitcoin enables peer-to-peer transactions without intermediaries, ensuring that the currency remains a communal asset rather than the purview of a privileged few.

Another foundational stone in the edifice of Bitcoin is trustlessness. In traditional finance, trust is a currency in and of itself; we put our faith in banks to manage our funds and execute transactions faithfully. Bitcoin, however, operates on a different philosophy—one where trust is replaced by verifiable, mathematical proof. The network uses complex cryptographic algorithms to ensure that transactions are secure and that once bitcoins are spent, they cannot be spent again—a concept known as double-spend protection. This cryptographic proof underpins the security of the Bitcoin network, allowing strangers to exchange value with the assurance that the system itself is impervious to duplicity.

Beyond decentralization and trustlessness, Bitcoin is also anchored by principles of anonymity and transparency— an ostensibly paradoxical pairing that is elegantly reconciled within the blockchain. While the identity of participants is shielded by pseudonymous addresses, ensuring a degree of privacy and security, the transactions themselves are recorded on a transparent ledger, open to all. This ledger, once written, becomes immutable, creating an indisputable record of transactions.

Additionally, the supply of Bitcoin is governed by an anti-inflationary ethos, a stark contrast to the modus operandi of fiat currencies. Where traditional money can be printed in limitless quantities at the behest of central banks, Bitcoin boasts a fixed supply cap of 21 million coins. This scarcity is coded into the very DNA of Bitcoin, providing a predictable monetary policy that stands in opposition to the inflationary tendencies of contemporary currencies. Moreover, Bitcoin's design allows each unit to be divisible down to its tiniest fraction, known as a 'satoshi' or 'sats' for short, named in homage to its enigmatic creator. There are 100,000,000 satoshis in every single Bitcoin. This level of divisibility makes microtransactions possible with Bitcoin, even as the value per Bitcoin has grown substantially since its inception.

Each newly 'mined' Bitcoin requires computational work, thereby embedding value in the currency through energy and effort, a concept reminiscent of precious metals, which need labor to be extracted from the earth.

These principles coalesce to form the bedrock upon which Bitcoin is built, offering an alternative to those disenchanted with the vulnerabilities of traditional financial systems. They serve as the guiding lights for a new economic pathway that seeks not only to redefine money but also to re-envision its management and movement mechanisms.

The Early Days of Bitcoin

In the nascent moments of Bitcoin's history, the mining of the Genesis Block—Block 0—stands as a monumental milestone. This first block, mined by Satoshi Nakamoto, is emblematic, a symbol of inception for a decentralized ledger that would challenge the fabric of traditional finance. The Genesis Block is imbued with significance not only because it represented the beginning but also due to the embedded message within its code: "The Times 03/Jan/2009 Chancellor on brink of second bailout for banks." This headline from The Times newspaper served as both a timestamp and a politically charged statement, hinting at Bitcoin's creation as an alternative in the midst of global financial instability.

The act of mining the Genesis Block set into motion the Bitcoin protocol, a system that would grow from a solitary block on a lone computer to a vast network sprawling across the globe. The implications of this first block were profound, serving as the immutable anchor point for all the blocks that would follow, each adding to the immutable chain and contributing to the network's overall security.

The early days of Bitcoin were also marked by its first transaction. Satoshi Nakamoto sent ten bitcoins to computer scientist Hal Finney, marking the first-ever transfer of bitcoin between individuals. This event was pivotal—it demonstrated the functionality of the network and set a precedent for the digital currency's potential. Bitcoin's value at that time was more of a concept than a market reality. Still, this exchange was the initial spark that would ignite a global conversation about the worth and potential of this new form of money.

As Bitcoin began to function as a medium of exchange, even if only on a small scale, the community around it started to develop. Early adopters, intrigued by the prospects of a decentralized currency and the technology underpinning it, gathered in forums and chat rooms, contributing to the development of the network. They were technologists, libertarians, cryptographers, and visionaries, each drawn to Bitcoin for various reasons, from its technical elegance to its philosophical implications.

These early community members were essential in sustaining and improving the network and fostering the grassroots spread of Bitcoin. They set up exchanges, maintained nodes, and, importantly, started the conversations that would bring Bitcoin out of the shadows of niche online forums and into the broader public consciousness. Their collective efforts laid the groundwork for a burgeoning ecosystem that would grow in complexity and scale, eventually capturing the world's attention.

The embryonic stages of Bitcoin were marked by significant challenges and milestones that would ultimately shape the cryptocurrency's trajectory. Among the most iconic of these early landmarks was the first recorded purchase of a tangible good with Bitcoin. In May 2010, a programmer named Laszlo Hanyecz made history by successfully trading 10,000 bitcoins for two delivered Papa John's pizzas. This transaction, celebrated annually in the Bitcoin community as "Bitcoin Pizza Day," was a crucial proof of concept. It demonstrated Bitcoin's potential as a means of real-world exchange. It established an initial, though informal, economic valuation for the nascent digital currency.

As the ecosystem evolved, establishing exchanges and marketplaces became the next critical phase in Bitcoin's development. These platforms facilitated the trading of bitcoins for traditional currencies, which helped establish a more formal market price and provided an entry point for new users. Creating such marketplaces was instrumental in moving Bitcoin from the periphery into the viewfinder of a larger audience, fostering liquidity and enabling the gradual accumulation of market capitalization.

> **Liquidity refers to how easily an asset or investment can be bought or sold in the market without significantly impacting its price.**

However, growth came with growing pains. The Bitcoin protocol, while revolutionary, was not without its need for refinement. The community faced the challenge of ensuring the network's security, scalability, and efficiency. Significant updates and protocol forks—alterations to the code that could be minor or sometimes result in a split into two separate blockchains—became subjects of heated debate within the community. These forks were critical in that they allowed the protocol to evolve and adapt to the growing needs of the network but also highlighted the challenges of governance in a decentralized system.

One such fork led to the creation of Bitcoin Cash, a new cryptocurrency born from a disagreement on how to scale the network. Other forks and updates aimed to improve aspects like the network's capacity to handle transactions, enhance privacy, or introduce new features like smart contract functionality. Each update served as a milestone in its own right, a testament to the ongoing process of consensus-building and the community's commitment to stewarding the network through its formative years.

Through these early challenges and milestones, Bitcoin demonstrated a resilience that lent credibility to the cryptocurrency. It solidified its place as a technological novelty and a viable, albeit fledgling, financial asset. The path could have been more straightforward, and the hurdles were many. Still, the milestones reached provided the building blocks for the robust platform Bitcoin would become.

Adoption & Growth

The growth of Bitcoin's adoption is a tale of how a groundbreaking idea transforms into a movement, thanks in large part to the conviction and contributions of its early adopters. These trailblazers, ranging from cryptography enthusiasts to libertarians, recognized Bitcoin's potential to upend traditional financial systems and provide an alternative means of conducting transactions. They were the miners who powered the network, the developers who refined the code, and the evangelists who spread the word. Their contributions were pivotal, not only in validating the network's functionality but also in laying the groundwork for the burgeoning community that would come to surround Bitcoin.

Among the early narratives that brought Bitcoin to a broader, albeit notorious, spotlight was the Silk Road—an online marketplace on the dark web. It operated from 2011 until its shutdown by the FBI in 2013. The Silk Road became synonymous with Bitcoin, as the cryptocurrency was used as the medium of exchange for various illicit goods and services. While this association with the dark web cast a shadow over Bitcoin's reputation, it also underscored the digital currency's underlying principle of censorship-resistant transactions. For all its infamy, the Silk Road inadvertently demonstrated one of Bitcoin's core tenets: the ability to transact freely without the need for traditional financial intermediaries.

During these early years, public perception and media coverage of Bitcoin was a volatile mix of intrigue, skepticism, and sensationalism. Bitcoin made headlines, often for extreme price volatility or its role in illicit online activities. However, these stories also brought Bitcoin into the public discourse, catalyzing discussions about the nature of money, privacy, and the role of state control in the financial sector. Each news cycle, whether positive or negative, increased the general awareness of Bitcoin, setting the stage for more mainstream adoption.

The media's portrayal of Bitcoin fluctuated as its price: It was variously hailed as the future of money by some and derided as a speculative bubble by others. As adoption grew, so did the network effect, which in turn drew the attention of more speculative investors, entrepreneurs looking to build on the Bitcoin ecosystem, and eventually, the curious gaze of the general public. This visibility, though sometimes controversial, was crucial in the transition of Bitcoin from an obscure digital phenomenon to a recognized financial asset, setting the foundation for businesses and institutions' next wave of adoption.

Regulatory Hurdles and Legal Recognition

As Bitcoin's adoption swelled and its presence on the global stage grew more prominent, it inevitably attracted the gaze of regulatory bodies and government entities. The initial wave of government responses to Bitcoin was a patchwork of reactions, ranging from curiosity and cautious acceptance to outright bans and warnings to consumers about the risks associated with digital currencies. Regulators grappled with classifying Bitcoin—was it a currency, a commodity, or something entirely new? The lack of a centralized issuing authority presented a problem for governments accustomed to regulating monetary systems through traditional financial institutions.

The journey toward legal recognition for Bitcoin was rife with legal challenges. In some jurisdictions, Bitcoin faced severe restrictions that stemmed from concerns over its potential for tax evasion, money laundering, and its use in criminal activities. The anonymity and borderless nature of transactions meant that existing legal frameworks were often inadequate to govern this new form of value exchange. Nevertheless, as the technology proved resilient and the community grew, a gradual path to recognition began to emerge. Some countries took the lead by establishing clear tax guidelines and instituting anti-money laundering standards for cryptocurrency transactions, thereby setting a precedent for other nations to emulate.

Internationally, the perspectives on Bitcoin's legitimacy varied greatly. While some nations embraced the innovation, seeing an opportunity to attract technological investment and foster financial inclusion, others maintained a staunch stance against using digital currencies. Across different continents, Bitcoin became a subject of intense discussion among policymakers and economic leaders, sparking debates about the future of money and national sovereignty over monetary policy.

These debates often revolved around a core dilemma: Should Bitcoin be integrated into the existing financial system, or should it remain an outlier, operating on the fringes of the economy? This question remains at the heart of international perspectives on Bitcoin's legitimacy. As some countries establish comprehensive, supportive laws that aim to integrate Bitcoin into their financial systems, others continue to resist, highlighting the diversity of approaches to this revolutionary technology.

In a bold stride toward financial innovation, El Salvador etched its name in the annals of economic history by adopting Bitcoin as legal tender, a global first that sent ripples through the monetary landscape. On June 9, 2021, under the leadership of President Nayib Bukele, the Salvadoran legislature passed a law that would integrate Bitcoin into the nation's financial infrastructure. This unprecedented move was rooted in a bid to boost economic growth, enhance financial inclusion for its unbanked population, and streamline the process of receiving remittances, which constitute a substantial portion of the country's GDP. The government introduced the 'Chivo' wallet, an application designed to facilitate Bitcoin transactions, promising ease of use and incentivizing adoption by crediting users with a starting amount of Bitcoin. Despite facing criticism and skepticism from traditional financial entities and grappling with the inherent volatility of the cryptocurrency, El Salvador's venture into Bitcoin represents a significant case study in the large-scale application of digital currency, marking a potential turning point in how nations perceive and engage with the concept of money in a digital age.

Despite the varied responses, one thing became clear: Bitcoin had sparked a global conversation about the nature of money, sovereignty, and the right to financial privacy. As these discussions evolved, so too did the legal and regulatory landscapes, adapting—sometimes reluctantly—to a world in which Bitcoin plays an increasingly significant role.

From the revolutionary manifesto embedded in its genesis block to the bustling ecosystem it has spawned, Bitcoin has ignited a financial and philosophical revolution. Its emergence as a reaction to the 2008 financial crisis represented not just the creation of a new currency but the birth of a radical idea—that money, like information, could be free from the control of centralized powers.

The evolution of Bitcoin from an obscure cryptographic experiment to a global phenomenon is nothing short of remarkable. It has transcended its initial status as the purview of tech enthusiasts and libertarians to command a presence on the world's economic stage. This digital currency, once known only in the narrow corridors of internet forums, now finds itself the subject of discussion among the world's leading financial institutions, governments, and an ever-growing community of users.

As we segue into the next chapter, we stand at the precipice of a new era. Bitcoin, once regarded as a novelty, an outsider in the realm of serious financial consideration, is now being woven into the very fabric of the financial sector. It challenges long-held assumptions about the nature of assets and the mechanisms of their exchange. This upcoming chapter will delve into Bitcoin's maturation, its integration into mainstream finance, and its emerging role as both a haven asset and a staple of diversified portfolios. We will explore how Bitcoin is redefining the boundaries between technology and economics, leading us into uncharted financial waters with the potential to reshape the future of money.

Chapter 6
Bitcoin: Digital Gold and a Store of Value

To contextualize Bitcoin within this framework, we must embark on a brief journey through history, exploring how various stores of value have underpinned and shaped economies. Throughout the ages, diverse assets have assumed this crucial role—from livestock and grains in ancient agricultural societies to precious metals like gold and silver in more recent history. Each has had its moment in the sun, epitomizing stability and security to those who held them.

The historical narrative brings us to the threshold of the modern era, where fiat currencies, backed by the full faith and credit of governments, have dominated. Yet, in the wake of financial crises and the advent of the digital age, a new player has entered the stage. Bitcoin, with its digital scarcity and borderless nature, presents a modern-day alternative, vying for inclusion in the pantheon of assets deemed worthy of the title 'store of value.' As we explore this proposition, we grapple with the idea that in an ever-evolving financial landscape, perhaps the most enduring store of value is one that adapts to the times.

As the curtain rises on the digital age, Bitcoin steps into the spotlight, asserting itself as a contemporary store of value. Its proponents herald its unique characteristics, which align closely with the attributes historically reserved for long-term value preservation. Bitcoin is divisible, portable, and, most notably, scarce—qualities that are imperative for any asset that aspires to maintain purchasing power over time. Its digital nature transcends physical borders, and its cryptographic foundation offers a level of security that is independent of traditional financial institutions.

In direct comparison to established stores of value, such as precious metals or real estate, Bitcoin offers a striking contrast. Unlike gold, there is no need for physical storage, and unlike real estate, it isn't tethered to a specific location. Its transferability surpasses that of bullion, and its divisibility exceeds that of most tangible assets, with one Bitcoin divisible down to 100 million satoshis. Furthermore, its digital existence means it can be transferred across the globe almost instantaneously—a feat no physical asset can claim.

The term 'Digital Gold' is thus often ascribed to Bitcoin, capturing the essence of its intended purpose. Gold has stood for centuries as a symbol of wealth and stability, a bastion against the fiat currencies' susceptibility to inflation and government manipulation. Bitcoin, in its digital sheen, takes this mantle into the cyber realm, proposing a new kind of asset that carries the old guard's reliability into the future of finance.

The emergence of Bitcoin as a store of value represents a fusion of historical precedence and innovative progression. As it continues to gain traction and acceptance, the narrative of Bitcoin is not just about its current status but its potential to redefine the very concept of what an enduring asset can be in an increasingly digitized world.

Bitcoin as a Hedge Against Inflation

Inflation is the silent thief of purchasing power, a relentless force that can erode the value of currency as the cost of goods and services inevitably rises. At the heart of inflationary mechanisms is the delicate balance between the money supply and the economy's capacity for goods and services. When a government prints more money or when the money supply increases faster than economic growth, the value of the currency tends to diminish. The more currency there is in circulation without a commensurate increase in economic output, the less each unit of currency can buy.

Bitcoin stands in stark opposition to inflation-prone fiat currencies with its inherently deflationary model. Encoded into its protocol is a finite limit of 21 million bitcoins that will ever exist. This scarcity is further reinforced by the halving event that occurs approximately every four years, reducing the rate at which new bitcoins are created. This built-in scarcity mimics the natural scarcity of precious resources like gold, but it does so in the digital realm, inherently limiting the possibility of devaluation through increased supply.

Moreover, Bitcoin's issuance is algorithmically determined, transparent, and predictable, removing the possibility of erratic changes in monetary policy that can lead to inflation. As such, Bitcoin is often posited as an inflation hedge — akin to digital gold — that can serve as a safeguard against the dilution of purchasing power. It offers a modern solution to an age-old problem, challenging the traditional fiat monetary systems that have, at times, led economies to the brink of collapse due to unchecked inflation.

Gold and Bitcoin: A Comparative Analysis

In the realm of value and wealth, the venerable gold and the insurgent Bitcoin stand as monuments to their times—gold, a glittering testament to tangible worth; Bitcoin, a cipher for the digital era's intangible promise. The contrast between the two is stark: gold, with its physical allure, has been a symbol of wealth for centuries, its permanence and heft conveying a sense of security. Bitcoin, by contrast, exists in the ether of cyberspace—a series of codes and consensuses that argue for a new kind of permanence.

Gold's tangible nature historically lent it an air of indisputable value—each bar, coin, or trinket could be touched, weighed, and admired. This physicality, however, is not without its burdens: gold requires secure storage, it must be protected, and its transportation is neither simple nor inconsequential. Bitcoin emerges as its antithesis in this regard, requiring no vaults for safekeeping, effortlessly crossing borders with the ease of data transmission, and its cryptographic keys a modern analog to gold's physical lock and key.

While both assets are perennially liquid, easily bought, and sold in sizeable markets worldwide, they diverge in their market behavior. Gold's price sails with the steady current of established financial tides, while Bitcoin rides the swells of innovation and speculation, making it the more volatile vessel. Yet, it is this very volatility that captures the attention of modern traders, its price fluctuations a siren song to those seeking rapid gains and willing to brave the corresponding risks.

The true test of an asset's mettle often comes in times of economic distress. Here, gold has a proven track record, serving as a haven for capital throughout tumultuous financial storms. Bitcoin's ledger, though far briefer, has begun to suggest its potential as a modern haven—its value proposition not yet thoroughly tested by time but increasingly viewed as a possible bulwark against the inflationary trends of fiat currencies.

In the ultimate analysis, the comparison between gold and Bitcoin is a reflection of the shifting landscapes of value storage. While gold carries the weight of history, Bitcoin carries the momentum of the future. In the future, the definition of what is considered a reliable store of value will continually expand. Both assets have their roles in a diversified portfolio, but their differences underscore the unique benefits and challenges they present to modern investors.

Fiat Currencies and Bitcoin: Divergent Paths

At the intersection of economics and technology, Bitcoin and fiat currencies embark on divergent paths. The fiat currency model, familiar and long-standing, is backed by government decree rather than physical commodities. While this model affords governments significant control over economic policy, it has its vulnerabilities. The very nature of fiat currency, unhinged from any tangible asset, leaves it susceptible to devaluation through inflation, which can be precipitated by the often politicized act of printing more money. The history of fiat is littered with tales of currencies that fell victim to such inflationary pressures, their purchasing power diluted as the money supply outstripped economic growth.

These inflationary pressures remain a pressing concern in modern economies, where the delicate balance of monetary supply and value is a constant juggling act for central banks. In times of economic distress, the temptation to print money can lead to significant long-term consequences, eroding wealth and skewing economic incentives. The very trust that underpins fiat currencies—their acceptance as a medium of exchange and store of value—can be eroded if that balance is mismanaged, leading to cycles of boom and bust that have characterized financial history.

Bitcoin emerges as a radical alternative to this well-trodden path. Its decentralized ledger, the blockchain, operates independently of any central authority. This decoupling from government-controlled money systems is not simply a technical feat; it represents a fundamental shift in how money can work. Bitcoin is not subject to the whims of policymakers or the economic priorities of any single nation. Instead, its supply is algorithmically capped, a policy set in digital stone that promises a predictable monetary policy free from the risk of runaway inflation.

This divergence between the paths of fiat currencies and Bitcoin could not be more profound. One path is paved with government backing and the flexibility (and potential for misuse) that such a system allows. The other forges its way through uncharted territory, with mathematical certainty as its compass and a global, borderless network as its territory. As we explore the ramifications of these divergent paths, it becomes clear that Bitcoin's emergence is not just a challenge to the fiat currency model; it is an invitation to reimagine what money can be in the age of information.

The Role of Scarcity in Valuing Bitcoin

In the world of economics, scarcity is a fundamental principle that dictates the value of a resource. It's a simple equation: the scarcer an asset, the more valuable it tends to be, assuming demand is present. This concept applies as much to the realms of precious metals and fine art as it does to the domain of cryptocurrencies. Scarcity can be manufactured, as with diamonds or limited edition items, or it can be a natural occurrence, as with land or vintage wines. It is this scarcity that has historically underpinned much of the value attributed to gold, turning it into a universally coveted treasure.

Enter Bitcoin, a digital asset that has ingeniously applied the principle of scarcity within the digital landscape— a space where replication is usually straightforward and abundant. At the heart of Bitcoin's valuation is its supply limit: a hard cap of 21 million bitcoins that will ever exist. This cap is encoded into Bitcoin's very algorithm, and its adherence is maintained by a global network of nodes running the Bitcoin software. Unlike fiat currencies, which can be printed ad infinitum, Bitcoin's maximum supply is immutable, providing a safeguard against the inflation that plagues traditional monetary systems.

Further reinforcing Bitcoin's scarcity is the process known as mining. Mining is the act of validating and recording transactions on the blockchain, a task performed by a decentralized army of computers competing to solve complex cryptographic puzzles. The difficulty of these puzzles adjusts over time, ensuring that the rate of Bitcoin creation remains steady despite increasing computational power or the number of miners in the network. This difficulty adjustment is crucial—it enforces the schedule of supply and ensures that the introduction of new bitcoins follows a predictable deceleration.

Scarcity as a driver of value is thus deeply embedded in the fabric of Bitcoin. It's not merely an academic concept but a palpable force that shapes the perception and valuation of this digital asset. As we witness Bitcoin's continued adoption and integration into financial markets, its scarcity becomes ever more critical. It offers a stark contrast to the elastic supply models of fiat currencies and positions Bitcoin as a unique asset that stands outside the traditional monetary paradigms. In the chapters to come, we will delve deeper into how this scarcity, combined with other intrinsic properties of Bitcoin, contributes to its emerging status as a new form of digital gold and a store of value for the modern age.

Individuals from every walk of life, drawn by the allure of its potential to both preserve and increase wealth, have started incorporating Bitcoin into their financial strategies. Not merely confined to the adventurous investor, Bitcoin has captivated the attention of diverse entities—ranging from pioneering tech firms that see kinship in its digital essence to traditional businesses hedging against economic uncertainty. These entities are not just passive spectators but active participants, bolstering Bitcoin's stature through their investments and, in doing so, endorsing a new era of asset diversification.

In regions battered by economic instability, where national currencies waver under the strain of hyperinflation and political strife, Bitcoin has emerged as a bulwark against the tide. For citizens of countries like Venezuela and Zimbabwe, where the local currency's purchasing power can evaporate in a matter of days, Bitcoin offers a semblance of stability and hope. Its immunity to unilateral devaluation by any government provides a semblance of financial sovereignty, often for the first time. The utility of Bitcoin goes beyond mere investment; it becomes a lifeline, a means to engage with the global economy on a more stable footing.

The most telling sign of Bitcoin's maturation as a store of value is its burgeoning relationship with institutional investors. Once skeptical, the guardians of multibillion-dollar funds are now charting a course into Bitcoin's waters. This shift is significant, a nod to Bitcoin's growing legitimacy and potential for longevity. Major financial institutions, endowments, and even publicly traded companies are not just dipping their toes but are increasingly wading deeper into the Bitcoin pool, carving out substantial allocations within their portfolios. Their engagement with Bitcoin reflects a profound recognition—crypto assets now form a critical component of a well-rounded investment strategy, offering a hedge against the inflationary trends of traditional fiat currencies.

Critics question Bitcoin's viability as a store of value, pointing to its volatility and the developing nature of the market. Skeptics argue that something so new and fluctuating, prone to dramatic price swings, cannot be compared to established stores of value like gold or real estate. They posit that Bitcoin's worth is interwoven with speculative trading and market sentiment, asserting that its value is not intrinsic but derived from collective belief and the willingness of the next buyer to pay a higher price.

Addressing such criticisms requires a foray into the heart of Bitcoin's value proposition. Its proponents contend that volatility is a natural phase in the life cycle of an emerging asset class, one that will stabilize with greater adoption and deeper market liquidity. They argue that the traditional stores of value, too, were once subject to the same doubts and volatility that Bitcoin faces today.

The concern over Bitcoin's energy consumption and its environmental impact forms another pillar of critique. The proof-of-work mechanism that underpins Bitcoin's security and trustlessness is energy-intensive by design, requiring a significant amount of computational power. Critics highlight the sustainability issue this presents, stressing the growing need for green energy solutions. In response, advocates point to the rapid advancements in renewable energy sources within the mining sector and the potential for Bitcoin's energy consumption to drive innovation in cleaner energy technologies.

Additionally, security concerns are often raised, with incidents of exchange hacks and lost private keys underlining the perceived risks. Yet, supporters of Bitcoin note that these are not flaws of the blockchain or the Bitcoin protocol but rather issues with third-party services and user practices. They underscore the importance of robust security measures, such as hardware wallets and multi-factor authentication, as the antidote to such vulnerabilities.

Conclusion

As we stand at the threshold of a new financial paradigm, the future outlook for Bitcoin as a store of value remains a subject of intense speculation and interest. It is a vista brimming with possibilities, where Bitcoin's narrative is as much about potential as it is about its current stature. Looking ahead, Bitcoin promises a vision of the financial future that is both exhilarating and, for some, disconcerting. Both milestones and obstacles mark its ascent to a recognized store of value, each new peak bringing with it broader acceptance and deeper integration into the global economy.

The burgeoning presence of Bitcoin has begun to influence shifts in asset allocation and investment strategy, nudging the needle towards a more diversified and, perhaps, more resilient approach to wealth preservation. Institutional investors, once bystanders, are now actively considering and, in many cases, already incorporating Bitcoin into their portfolios, signaling a seismic shift in investment philosophy. As more entities adopt this digital asset, we may witness a reshaping of the investment landscape, with Bitcoin emerging as a fundamental component alongside traditional investments.

Bitcoin's role in the global economy is evolving from that of an outsider to a force majeure, one that commands attention and demands consideration. It is a catalyst for change, challenging entrenched financial doctrines and offering an alternative narrative to the story of money. The implications of Bitcoin's ascent extend beyond mere market dynamics, touching on broader themes of autonomy, sovereignty, and the democratization of finance.

As we approach the horizon where the current financial ethos intersects with the digital revolution, the next chapter promises to be an expedition into the intricate lattice that is Bitcoin's technical infrastructure. We will demystify the complex algorithms and the cryptographic wizardry that form the backbone of Bitcoin, delving into the mechanics of the blockchain, mining, and the genius of decentralized consensus. This exploration will unveil the meticulous engineering that ensures Bitcoin's security and reliability, allowing it to stand as a paragon of digital currency.

Anticipation builds as we consider how this technological bedrock supports Bitcoin's lofty value proposition. In the forthcoming chapter, we will dissect the protocol layer by layer, revealing how each component contributes to the overarching narrative of Bitcoin as a store of value. How does this digital alchemy transform strings of code into digital gold? What makes the protocol resilient against the assaults of hackers and the fluctuations of market sentiment? These are the questions that will guide our inquiry.

Prepare to embark on a journey through the digital veins of Bitcoin as we unveil the secrets that govern its heart and sustain its pulse. The next chapter is not just a technical overview; it is a story of innovation and the relentless pursuit of a system that reimagines the very concept of value in an interconnected world.

Section III
Understanding Bitcoin

Chapter 7
The Bitcoin Protocol: How It Works

As we embark on this detailed exploration of the Bitcoin protocol, it's important to note that a thorough understanding of the technical mechanics described in this chapter is not a prerequisite for using Bitcoin. Much like one doesn't need to be an automotive engineer to drive a car, you don't need to be a blockchain expert to use Bitcoin. The upcoming exposition is designed for those curious minds who yearn to peer under the hood, to truly grasp the inner workings of this digital currency. For the everyday user, the Bitcoin network offers a user-friendly interface through various wallets and services that make transacting with Bitcoin as simple as sending an email. So, whether you're here to simply learn the basics of using Bitcoin or to dive deep into the complexities of its infrastructure, this chapter will serve as a guide to enhance your understanding of one of the most revolutionary technologies of our time.

At its core, the Bitcoin protocol is a marvel of modern technology—a digital alchemy that transforms intricate mathematical and cryptographic principles into a secure and functional currency. Its foundation is the blockchain, a decentralized ledger that chronicles every transaction ever made within the network. This ledger is maintained across a vast tapestry of nodes, ensuring that no central point of failure could compromise the currency's integrity. The Bitcoin protocol operates on the principles of cryptography, which secure transactions and provide the framework for the creation and transfer of the currency itself, known colloquially as bitcoins.

The purpose of the Bitcoin protocol extends beyond the mere creation of digital coins. It is an embodiment of a decentralized financial system, free from the control of any single institution or government. Its functions are manifold, facilitating secure peer-to-peer transactions, ensuring the fidelity of the currency's supply, and allowing participants, known as miners, to maintain the network's operations. These miners are rewarded for their efforts in new bitcoins, a process that not only upholds the network's security but also systematically introduces new currency into the system, without the need for a central mint.

As we delve further into the inner workings of the Bitcoin protocol, remember that it stands as the backbone of an entirely new form of money—one that is not printed, but 'mined'; not centralized, but distributed; and not held in vaults, but in digital wallets. This introductory glimpse sets the stage for a more in-depth journey through the protocol's intricacies, revealing how Bitcoin is engineered to function as a secure, independent, and borderless currency in the digital age.

The Blockchain: Bitcoin's Public Ledger

Envision a ledger, not bound by the physical constraints of paper and ink, but existing in the digital expanse—a ledger that is public, continuously verified, and immutable. This is the blockchain, the foundational technology underpinning Bitcoin. It is a sequence of blocks, each a repository of transaction data, chained together by cryptographic links that render the ledger tamper-evident. The blockchain's architecture eschews traditional centralized record-keeping in favor of distributed consensus, ensuring that each transaction is an indelible part of financial history.

When a Bitcoin transaction is made, it doesn't become official immediately. Instead, it joins a pool of other transactions in a state of limbo, awaiting confirmation. Miners—participants in the network—gather these transactions and form them into a block. Through the process of mining, which involves solving a complex cryptographic puzzle, the miner seeks the network's approval to add the new block to the chain. Once added, the transaction is considered confirmed, the funds securely transferred, and the information irrevocably etched into the blockchain.

The blockchain's role in the Bitcoin ecosystem is one of unwavering guardianship. It is the standard-bearer of integrity and trust, eliminating the need for intermediaries such as banks or clearinghouses. Trust in the system is not placed in a single entity but is instead diffused across the network—a network that collectively agrees on the validity of each transaction and the state of the ledger at any given moment. This decentralized trust model is revolutionary, redefining the principles of financial security and opening up a world where the exchange of value is transparent, efficient, and accessible to all.

Wallet Addresses and Keys

The gateway to interacting with Bitcoin is the creation of a digital wallet, an action akin to opening a bank account in the traditional financial world. However, unlike opening a bank account, creating a Bitcoin wallet doesn't require a visit to a financial institution or any paperwork; it can be done by anyone, anywhere, using random number generation.

This process begins with the generation of a private key, a number so vast in possibilities that the chance of randomly generating a key already in use is nearly infinitesimal. It's a number that resides in the realm of 2^256 possibilities when using the Bitcoin protocol, a space so immense that it dwarfs the estimated number of atoms in the observable universe. This staggering level of randomness is what gives each Bitcoin wallet its security.

Indeed, anyone can create a wallet independently, and there are a variety of methods to do so, ensuring that one's method of creation can be tailored to one's comfort with technology and security. One such method is the use of physical dice. By rolling dice and using the results to contribute to entropy or randomness, users can manually generate a private key without the need for a computer, known as the 'dice method.' This approach, while more labor-intensive, is praised for its removal of reliance on potentially compromised computer systems and offers the user a tangible hand in the creation of their digital wallet.

Once a private key has been generated, it is then used to produce a public key through cryptographic functions. From this public key, a wallet address is derived—a string of characters that represents the destination for sending and receiving Bitcoin. This address can be shared openly, functioning similarly to an email address in terms of privacy, whereas the private key must be guarded with utmost secrecy as it is the tool that grants access to send the Bitcoin held in that wallet.

This dichotomy of public and private keys is at the heart of Bitcoin's security model. It empowers individuals to take control of their finances with a level of privacy and security that is not predicated on the identity or location of the user but solely on their ability to safeguard their private key. It's a revolutionary shift from traditional financial systems, where identity and trust are central, to one where anonymity and cryptographic proof offer a new paradigm for financial autonomy.

Transactions: The Building Blocks of the Blockchain

At the heart of Bitcoin's blockchain are transactions, the fundamental units that denote the transfer of value from one party to another. Each transaction is a carefully constructed digital package of data that details the sender, the receiver, the amount of Bitcoin being transferred, and a set of digital signatures verifying the sender's intent. This anatomy is intricate, with each part playing a critical role in ensuring the transaction's legitimacy and security.

Central to these transactions are wallet addresses and keys. A wallet address functions much like an account number—a unique identifier that marks the destination or origin of Bitcoin in a transaction. These addresses are derived from public keys, part of a cryptographic key pair that each wallet holder possesses. The counterpart to the public key is the private key, which is akin to a personal digital signature or PIN. It is this private key that provides the authority to send bitcoins from a particular wallet, ensuring that only the rightful owner can initiate transactions.

Delving deeper into the transaction structure, we find that each transaction consists of inputs and outputs. Inputs refer to the origins of the bitcoins being sent, essentially referencing where the sender received their bitcoins from, typically from a previous transaction output. The outputs then designate the next destination for those bitcoins, which can be another user's wallet address or, in the case of transactions that aren't sending the entire balance, a return address to the sender's wallet. It's this flow of inputs and outputs that knits together the whole transaction history of Bitcoin on the blockchain, creating a seamless tapestry of digital value exchange.

Understanding the mechanisms of Bitcoin transactions is crucial for grasping the larger workings of the blockchain. These transactions are the sinews that connect the network's users, allowing them to engage in secure digital exchanges without the need for intermediary validation. They embody the essence of Bitcoin's decentralized ethos, where trust is established not through institutions but through cryptography and the collective agreement of the network's participants.

Scripting and Smart Contracts

Bitcoin's underlying capabilities are bolstered by a scripting language that permits the formation of conditions necessary for the expenditure of bitcoins. This scripting language, deliberately constrained to be non-Turing complete, which means it doesn't have the same capabilities as a full programming language. Its instructions, or opcodes, are executed in a simple stack-based processing system. It's a design that ensures predictability and safeguards against the computational perils of loops, thereby bolstering the overall resilience of the network. Most commonly, these scripts dictate that bitcoins can only be transferred with the proper authorization, typically a digital signature verified against the public key associated with the bitcoins.

The utility of Bitcoin's scripting extends to enable a

variety of transaction types. It underpins the functionality of multi-signature transactions that necessitate multiple approvals before funds can be disbursed, providing enhanced security. Additionally, scripts can encode conditions based on time, such as transactions that only become valid after a certain block number or date—useful for escrow or deferred payments. These capabilities represent the rudimentary building blocks of smart contracts within the Bitcoin ecosystem—a way to set predetermined conditions for the movement of bitcoins.

Despite these functional forays, Bitcoin's scripting language does not boast the same range of possibilities as those found on more complex smart contract platforms like Ethereum. Its simplicity is a feature, not a flaw, ensuring the integrity of the system at the expense of wider functionality. This has, understandably, placed some constraints on the scope of decentralized applications that can be built on Bitcoin. However, innovation within the Bitcoin community continues, with improvements like the Taproot upgrade, which aims to enhance the functionality and efficiency of these scripts.

Developments are also being made in the form of sidechains and layered solutions like the Lightning Network, which propose to extend the capabilities of Bitcoin's scripting without overhauling the core protocol. These initiatives hint at a potential future where Bitcoin's functionality could be expanded, allowing for a broader array of financial instruments and services to be built atop this most secure and established blockchain. Bitcoin's scripting language is, therefore, a dynamic element of its protocol—a feature that is simultaneously a bastion of stability and a seedbed of potential growth.

Mining: The Engine of Network Security

When this book uses the term "cryptographic puzzles," it is referring to a specific and vital process. This process is not simply about creating new bitcoins or processing transactions; it is integral to the security and functionality of the Bitcoin network. At its core, mining involves constructing a block of transactions and discovering a specific magic number known as the nonce, which stands for "number only used once." This nonce, when hashed along with the block data, must produce a hash value that is lower than or equal to the network's current difficulty target.

The difficulty target is a critical component of the network, set to ensure the stability and security of Bitcoin's transaction processing. It adjusts periodically to maintain an average block time, which is typically around ten minutes. This adjustment is necessary to account for changes in the network's total computational power.

When miners attempt to solve the "cryptographic puzzle," they engage in a trial-and-error process to find a nonce value that meets this criteria. They continually hash the block data with different nonce values until they find one that produces a hash value lower than the difficulty target. The first miner to achieve this successfully gets to add the new block to the blockchain, thereby confirming and securing the transactions within it.

The successful miner is rewarded for their efforts with new bitcoins—the block reward—and the transaction fees from all transactions included in that block. This reward system incentivizes miners to contribute their computational power to the network, ensuring a decentralized and secure transaction validation process.

In essence, the "cryptographic puzzles" in Bitcoin mining are a sophisticated and ingenious mechanism. They ensure the integrity of the blockchain, regulate the introduction of new bitcoins, and underpin the decentralized nature of the Bitcoin network. The creation of new blocks through mining is a competitive and resource-intensive endeavor, but it is one of the ingenious innovations that make Bitcoin a secure and decentralized currency. Every ten minutes, on average, the history of Bitcoin's transactions is sealed in a new block, added to a chain that is a testament to the combined effort and integrity of the network's miners. It is through mining that Bitcoin transcends being merely a digital currency and becomes a continuously affirmed social contract underpinned by the principles of cryptography and decentralized consensus.

Nodes: The Network's Watchtowers

In the vast, distributed ledger that is the Bitcoin network, nodes serve as the vigilant custodians of its history and the judges of its truth. Each node in the network is a guardian, running a copy of the Bitcoin software, which contains the entire history of transactions within the blockchain. They function as the individual cells of a larger organism, collectively maintaining the health and integrity of the system. Nodes enforce the rules of Bitcoin by refusing to accept or propagate transactions and blocks that violate the protocol's standards.

The role of nodes is crucial in the validation process. Whenever a new transaction is broadcast to the network, it is the nodes that shoulder the responsibility of ensuring its validity. They scrutinize each transaction against Bitcoin's stringent consensus rules; if a transaction fails to comply, it is rejected, never to be inscribed in the sacred ledger. Once validated, these transactions await inclusion in a new block. When a miner successfully adds a block to the blockchain, nodes perform a critical final check, verifying the block's adherence to the protocol before cementing it into their version of the blockchain.

The Bitcoin network is composed of various types of nodes, each contributing differently to the ecosystem. Full nodes hold a complete copy of the blockchain and fully enforce all the rules of the Bitcoin protocol. They serve as the ultimate gatekeepers, providing the highest level of security and autonomy in transaction verification.On the other hand, lightweight or "SPV" (Simplified Payment Verification) nodes offer a way to verify transactions without having to download the entire blockchain and rely on the data provided by full nodes to interact with the network. Mining nodes, often coupled with full nodes, add new transactions to the block they're looking to mine in the hopes of solving the computational puzzle that will allow them to add a new block to the chain. There are also listening nodes, sometimes referred to as 'relay' nodes, which play a critical role in the propagation of information across the network.

Together, these nodes form an intricate web of oversight and communication. They are the bedrock of Bitcoin's decentralized nature, ensuring that no single point of control can compromise the fidelity of the blockchain. The system's trustworthiness and security derive from this decentralized architecture, where consensus is not a product of authority but of a collective agreement and adherence to the immutable laws of mathematics and cryptography encoded within the Bitcoin protocol.

Within the sprawling digital tapestry that constitutes the Bitcoin network, a multitude of individuals and organizations serve as the custodians of its distributed ledger, operating nodes for various reasons. Individual enthusiasts champion the decentralized and privacy-centric ethos of Bitcoin, running nodes to strengthen the network and enhance the security and anonymity. Miners, the linchpins of Bitcoin's transaction validation and block creation process, inherently require nodes to partake in and contribute to the network's continuity and economic incentives.

Businesses entrenched in the Bitcoin ecosystem, such as exchanges and wallet services, operate nodes as the sinews connecting their operations to the blockchain, ensuring transactional accuracy and autonomy. For them, nodes are vital tools that support service reliability and customer trust. Simultaneously, developers and entrepreneurs, ever on the quest for innovation within the space, maintain nodes to test new applications and ideas, pushing the boundaries of what the network can achieve.

Academic and research institutions often operate nodes for the purpose of conducting practical and educational research into this financial phenomenon. Running a node provides valuable insights and offers hands-on experience in the field of cryptocurrency. This approach allows scholars and students alike to deeply understand the mechanics and technology behind cryptocurrencies. By engaging directly with the blockchain network through their own node, these institutions can study its workings in real-time, offering a dynamic learning experience that goes beyond theoretical knowledge. This practical involvement is crucial in helping the academic community keep pace with the rapidly evolving world of digital finance and blockchain technology

The motivations behind node operation are as varied as the participants themselves. Some are driven by ideological alignment with Bitcoin's vision of a trustless and open financial system. Others see it as an investment in the network's robustness, correlating the health of this ecosystem with the potential appreciation of Bitcoin's value. Then there are those for whom running a node represents a pursuit of independence, a way to engage with the cryptocurrency realm on their own terms, relying on no one but the code itself.

Together, these individuals and entities form the backbone of Bitcoin, upholding the decentralization that makes the network so resilient. They are the silent sentinels who, without fanfare, ensure that Bitcoin remains true to its foundational principles of openness, security, and peer-to-peer exchange of value. Through their collective efforts, they embody the spirit of the protocol and safeguard the ledger that is redefining the landscape of money and finance.

Consensus Mechanisms: Establishing Network Agreement

In the orchestration of the Bitcoin network, the chorus of consensus is a symphony that plays continuously, an essential harmony that maintains the ledger's integrity and trust. Consensus in Bitcoin is the process by which all the network participants agree on the state of the ledger, ensuring that each transaction is only confirmed once and that the shared truth is upheld across the global system. This agreement is not achieved through the edict of a central authority but through a distributed protocol—the bedrock of Bitcoin's democratic and trustless nature.

At the heart of this process is the Proof of Work (PoW) mechanism, an ingenious solution to the digital consensus problem. PoW is a cryptographic puzzle, a digital contest where miners across the world compete to solve complex mathematical problems. The first to arrive at the solution earns the right to add the next block of transactions to the blockchain and is rewarded with newly minted bitcoins. This reward system incentivizes the vast expenditure of computational resources required to maintain the network's operation, a testament to the security and reliability of the blockchain.

The beauty of PoW lies in its simplicity and effectiveness in preserving Bitcoin's decentralized ethos. It mitigates the risk of any single party monopolizing the network's transactional history, as the computational work required to influence the blockchain is prohibitively high. By tying the probability of mining a new block to the amount of computational work done rather than the miner's identity or status, PoW anchors the network's security in the laws of physics rather than trust in fallible institutions. The result is a system where trust is spread out across a vast array of independent nodes, each checking and verifying the others' work, making Bitcoin not just a ledger of transactions but a ledger of collective trust built on the foundation of consensus through Proof of Work.

Network Propagation

The Bitcoin network, a complex web of interconnected nodes, thrives on the rapid and reliable propagation of information—a process that is crucial for maintaining the ledger's consistency and security. When a transaction is broadcast to the network, it ripples outwards, node by node, like a pebble thrown into a digital pond. Each node that receives the transaction verifies it against the network's rules and then passes it on to its peers. This decentralized method of dissemination ensures that the transaction reaches most of the network within seconds, a testament to the system's design, emphasizing speed and efficiency.

This propagation is not just about speed but also about the network's ability to stay synchronized. Despite the vastness of the global Bitcoin network and the differing technological capabilities of its participants, the blockchain remains coherent. Each node maintains its own copy of the blockchain, which is updated with new transactions as they are confirmed in newly mined blocks. Through the inherent redundancy of its design, the Bitcoin network is robust against attempts to falsify transaction history, ensuring resiliency against attacks and network partitions.

In essence, the propagation of information across the Bitcoin network is a delicate balance of broadcasting data widely and maintaining a singular version of truth in a decentralized manner. It's a system that defies traditional single points of failure, is robust in its distribution, and resolute in its synchronization—qualities that make Bitcoin not just a cryptocurrency but a pioneering model for the future of decentralized digital systems.

Security Measures in the Bitcoin Protocol

The Bitcoin protocol is a fortress built upon the bedrock of cryptographic techniques, which secure every facet of its operation. At its core is the use of hash functions, particularly SHA-256, which turns data into a fixed-size hash—a string of characters that appears random but is deterministically produced from the input data. It's this hashing that secures the blocks and makes altering the blockchain practically impossible without detection. Public key cryptography is another pillar, enabling users to generate pairs of keys that facilitate secure Bitcoin transactions. The public key can be shared and used to receive funds, while the private key, kept secret, is used to sign transactions, providing proof of ownership digitally.

In order to validate transactions, the network employs several security protocols that verify the authenticity and integrity of each transaction. Each transaction broadcast to the network must be a valid structure and correctly signed as per the elliptic curve digital signature algorithm (ECDSA) to be considered valid. This cryptographic signature proves that the transaction was created by the rightful owner of the bitcoins in question, preventing unauthorized spending.

Moreover, the Bitcoin protocol incorporates robust measures against double-spending and fraud. Nodes in the network swiftly propagate transaction information, and once a transaction is confirmed and included in a block on the blockchain, the global ledger is updated. Any attempt at double-spending would require an adversary to outpace the entire network's computational power and mine an alternative chain of blocks. This feat becomes exponentially more difficult with every subsequent block added after the transaction. Additionally, the decentralized consensus mechanism ensures that for a fraudulent transaction to be accepted, an attacker would need to control more than half of the network's mining power, a scenario known as a 51% attack, which is highly improbable due to the vast and distributed nature of the network's hash rate.

These cryptographic foundations and security measures

work in concert to create a system that is remarkably resistant to fraud and unauthorized manipulation. The interlocking web of cryptographic assurances provided by the Bitcoin protocol not only secures individual transactions but also maintains the sanctity of the Bitcoin network as a whole, ensuring that it remains a trusted and secure platform for digital value exchange.

Forks and Protocol Variants

In the dynamic world of Bitcoin, 'forks' are events where the protocol diverges into two separate paths, often leading to significant changes in the network's operation and occasionally giving birth to a new cryptocurrency altogether. These forks are categorized as either 'soft' or 'hard,' with each type carrying its own set of implications for the network's future.

Soft forks are upgrades that tighten or add new rules to the blockchain protocol and are backward-compatible. They don't require all nodes to update, as the new rules still fit within the old rule set. This means that non-updated nodes will still see the blocks produced after the soft fork as valid. However, if the majority of the mining power does not enforce the new rules, this can lead to security vulnerabilities.

Hard forks, in contrast, are not backward-compatible. They introduce changes to the protocol that create permanent divergence from the previous version. All nodes must upgrade to the new protocol to continue participating in the network. If consensus on the upgrade isn't reached, it may result in two persisting blockchain versions, effectively splitting the community and the cryptocurrency into two.

Historical forks within Bitcoin have had a range of impacts. Notably, a hard fork in August 2017 resulted in the creation of Bitcoin Cash, a new cryptocurrency that implemented an increased block size to allow more transactions per block. This fork was the result of a long-standing debate within the Bitcoin community over how best to scale the network. Such events are significant, as they reflect the decentralized decision-making process and the varying philosophies within the Bitcoin community regarding the currency's future direction.

The implications of forks for users and developers are multifaceted. Users must be aware of forks as they can affect the security and value of their holdings. For instance, in the case of a hard fork, users may end up with coins on both versions of the blockchain, which can be both an opportunity and a risk. Developers, on the other hand, must consider forks when building applications or services that rely on the Bitcoin protocol, as they might need to ensure compatibility with the changes or decide which version of the protocol to support.

Forks represent a form of evolution within the Bitcoin ecosystem—a way for the protocol to adapt and grow in response to its users' collective needs and beliefs. While they can be contentious, they also underscore the fundamentally democratic and open-source nature of Bitcoin's development. Each fork is a crossroads, a decision point where the community's collective will shape the path Bitcoin will take into the future.

If a government attempted to impose an unpopular fork on the Bitcoin network, such as changing its maximum supply, the endeavor would likely be met with resistance from the community that upholds the network. Bitcoin's decentralized nature requires consensus for alterations to the protocol, and an arbitrary increase to the 21 million bitcoin limit would fundamentally conflict with the community's principles. In the face of such a proposal, most miners and node operators, who maintain the backbone of the network, would probably continue supporting the original protocol, preserving the sanctity of the established supply cap.

Such a unilateral move by a government would likely result in the creation of an entirely new cryptocurrency, while the original Bitcoin, with its adhered-to protocol, would persist as long as the community's support remained steadfast. Market participants would dictate the value of both the original Bitcoin and the new variant, and it's conceivable that the original Bitcoin would maintain greater value, reflecting the community's commitment to its foundational monetary policy.

A government-enforced alteration could significantly undermine the trust that is pivotal to the Bitcoin ecosystem. The original chain, revered for its adherence to the core tenets of fixed supply and decentralization, would likely retain its legitimacy and continue to be globally supported by developers and businesses within the space.

Moreover, such a forceful change could provoke legal disputes and political pushback, particularly from stakeholders viewing the intervention as an infringement on the autonomy of a decentralized financial system. The ramifications of a government-imposed fork extend beyond technology, striking at the heart of the ideological and economic pillars that support Bitcoin. Ultimately, the original Bitcoin protocol, resistant to coerced modification, would likely survive, a testament to the enduring power of a decentralized and collectively maintained network.

Transparency and Verification in Bitcoin Transactions

In the realm of Bitcoin, each transaction's journey from initiation to confirmation is a study of the balance of transparency and security. The verification of Bitcoin transactions is a meticulous process, hinging on the network's nodes. Each node scrutinizes incoming transactions against a rigorous set of criteria—confirming that the digital signatures are valid, the inputs have not been previously spent, and the parameters of the transaction adhere strictly to the network's protocol. This verification serves as the first line of defense against fraud, ensuring that only legitimate transactions are confirmed and added to the blockchain.

The architecture of Bitcoin ensures that transaction data is public, allowing anyone to view the transactions recorded on the blockchain. This transparency is fundamental to the trustless nature of Bitcoin; it allows for independent verification without the need for intermediaries. Every transaction, once confirmed, becomes a permanent part of the blockchain, accessible to anyone who wishes to explore the transactional history of the network. This degree of openness is unique to decentralized digital currencies and is emblematic of the paradigm shift that Bitcoin represents.

Yet, within this transparent system, the privacy of users is preserved through the use of pseudonyms—the wallet addresses. While all transaction details are public, the identities of the individuals conducting the transactions are not directly revealed. A wallet address appears as a string of alphanumeric characters, providing a layer of privacy for the users. Sophisticated techniques are required to trace transactions back to real-world identities. For those seeking additional privacy, further measures such as mixing services or privacy-focused wallets can obfuscate transactional trails.

This intricate dance of transparency and privacy is core to Bitcoin's design. It ensures that while everyone can trust the system as a whole, the individual user retains a degree of anonymity akin to that which is enjoyed in traditional cash transactions. The transparent yet pseudonymous nature of Bitcoin transactions upholds the integrity of the network while respecting the privacy of its participants, encapsulating the essence of a system built for an era that values both accountability and individual liberty.

Upgrades and Improvements to the Protocol

Bitcoin's protocol, a testament to resilience and adaptability, has seen its architecture refined and enhanced over time through a series of upgrades. These modifications showcase the currency's flexibility and the community's unwavering dedication to its evolution. Historical advancements such as Segregated Witness (SegWit) have expanded Bitcoin's transaction capacity and improved network scalability, reflecting a responsive system attuned to the growing demands of its users. More recently, the Taproot upgrade marked a significant leap forward, furthering transaction efficiency and privacy—a clear sign of Bitcoin's maturing technology.

Introduced in November 2021, Taproot represents a sophisticated development in Bitcoin's journey. It redefines transaction compactness and confidentiality through Merkelized Abstract Syntax Trees (MAST), offering a more discreet and concise method to set transaction conditions. This is particularly advantageous for complex transactions, such as those entwining multiple signatures or smart contracts, granting them a new stratum of privacy and streamlined efficiency.

A key innovation within Taproot is the optimization of transaction space, especially for those with numerous inputs and outputs. Prior to Taproot, the full details of a transaction's possible spending paths were transparent, but with the upgrade, only the executed condition was disclosed, simultaneously bolstering privacy and conserving block space. This leap not only camouflages the intricacies of complex transactions but also enhances the blockchain's ability to host more transactions.

Schnorr signatures, also incorporated within Taproot, replace the previous Elliptic Curve Digital Signature Algorithm (ECDSA). This new form of signature allows multiple signatures to merge into a single one on the blockchain, streamlining the verification process and further anonymizing transactions.

The process of refining Bitcoin through such upgrades is methodical and consensus-driven. Bitcoin Improvement Proposals (BIPs) are meticulously evaluated and tested before they are presented to the community for consideration. The implementation of changes, which may result in soft or hard forks, hinges on widespread consensus among miners and users, ensuring that the network's evolution is a collective progression rather than a top-down imposition.

As we peer into Bitcoin's horizon, the path of its protocol evolution stretches far and wide, with possibilities that ignite vigorous discussions within the community. Challenges in scalability, privacy, and efficiency beckon for solutions like the Lightning Network, which promises swifter and more cost-effective transactions without upending the core protocol. With the ever-shifting digital landscape and emergent technologies, Bitcoin's protocol is poised to adapt, shaped by the decentralized ethos that is the hallmark of this pioneering digital currency.

Chapter 8
Proof of Work Mining:
Block Rewards, Hash Rates, and
Energy Consumption

Understanding cryptocurrency mining requires more than a grasp of its technical intricacies. It's equally crucial to comprehend its economic implications. Mining is not just about the generation of new coins or processing transactions; it's also an energy-intensive process that has significant financial costs and environmental impacts. The delicate balance between mining costs such as electricity and hardware and rewards such as block rewards and transaction fees is constantly shifting. This balance influences miners' decisions, affects the overall network's security and efficiency, and has broader implications for the cryptocurrency market and energy consumption patterns worldwide.

Moreover, as cryptocurrencies continue to gain mainstream adoption, their mining processes attract more scrutiny and regulatory interest. The economic and technical aspects of mining are intertwined with market dynamics, technological advancements, and energy policies. Thus, a comprehensive understanding of cryptocurrency mining is not just beneficial for those directly involved in the mining process but also for investors, regulators, and anyone interested in the future landscape of digital currencies. This interplay of technology and economics not only shapes the present state of cryptocurrencies but also paves the way for their future evolution.

Delving into the fundamentals of cryptocurrency mining reveals a world where technology and economics converge in a unique dance. At the core of this process are key concepts that define its nature and impact: block rewards, hash rate, mining difficulty, and energy consumption.

Block rewards are the linchpin of the mining process. They represent the number of new cryptocurrency units awarded to miners for successfully validating a new block of transactions. This reward serves a dual purpose: incentivizing miners to contribute their computational power to the network and introducing new currency units into circulation, a process pivotal to cryptocurrencies that don't have a central issuing authority like traditional fiat currencies.

The hash rate is the total computational power being used to mine and process transactions on a blockchain network. It's a crucial metric reflecting the health and security of the network. A higher hash rate means more competition among miners and, by extension, a more secure network, as it becomes increasingly difficult for a single entity to gain control over the blockchain.

Mining difficulty is a dynamic aspect of cryptocurrency mining that adjusts in response to the network's total hash rate. As more miners join the network and the hash rate increases, the difficulty of solving the cryptographic puzzles also rises, maintaining a balance that ensures the steady creation of new blocks. This mechanism prevents the rapid mining of all available currency units and stabilizes the rate at which transactions are validated.

Energy consumption in cryptocurrency mining is a topic of significant interest and concern. The process is energy-intensive, as it requires a vast amount of computational power. The cost of electricity thus becomes a critical factor in the profitability of mining operations. Miners seek to optimize their hardware to maximize efficiency and reduce energy costs, a factor that has broader implications for environmental sustainability.

The actual process of mining involves miners using specialized hardware to solve complex mathematical puzzles. The first miner to solve the puzzle for each block is allowed to add that block to the blockchain and, in return, receives the block reward in the form of new cryptocurrency units and transaction fees paid by users. This incentivization model not only secures the network by validating transactions but also ensures that the process of introducing new currency units is decentralized and based on a competitive consensus mechanism.

Block rewards, which are essential in motivating miners to contribute their computational power to the network, have a significant impact on the economic incentives in cryptocurrency mining. These rewards, which consist of new cryptocurrency units and transaction fees, are the mining process's financial backbone. As miners successfully validate and add new blocks to the blockchain, they receive these rewards, making mining a potentially lucrative endeavor. This incentive is crucial for the security and functionality of the blockchain, as it encourages a decentralized network of miners to invest resources in maintaining and securing the ledger.

The impact of mining rewards on the hash rate is direct and significant. As the value of the rewards increases, either due to a rise in the price of the cryptocurrency or an increase in transaction fees, more miners are incentivized to join the network, thereby increasing the total hash rate. This increase in hash rate signifies greater security but also triggers an increase in mining difficulty, ensuring the stability of block creation times. Conversely, if the value of the rewards decreases, the hash rate may decline as miners with higher operational costs find it less profitable to continue mining.

The relationship between cryptocurrency market dynamics and mining activity is complex and intertwined. The price of a cryptocurrency can significantly influence mining profitability. When prices are high, mining becomes more attractive, drawing more participants and increasing the hash rate. This dynamic can create a feedback loop where rising prices lead to more mining activity, which in turn may positively influence market perceptions of the cryptocurrency's stability and security, potentially driving prices even higher.

Conversely, market downturns can reduce mining activity. Lower cryptocurrency prices mean reduced block rewards in fiat terms, leading to lower profitability for miners, especially those with higher electricity and operational costs. This situation can lead to a decline in the hash rate as less efficient miners exit the network. However, this also lowers the mining difficulty, potentially making mining viable again for the remaining participants and highlighting the self-regulating nature of blockchain networks.

The interplay between mining difficulty and hash rate is a crucial aspect of how cryptocurrency networks operate and maintain their security. To grasp this relationship, one must first understand the mining difficulty adjustment mechanism. This mechanism is integral to cryptocurrencies like Bitcoin and is designed to keep the rate of block creation consistent. It works by adjusting the complexity of the cryptographic puzzles that miners need to solve in order to validate transactions and create new blocks. This difficulty is recalibrated at regular intervals based on the total computational power, or hash rate, of the network. When the hash rate increases, indicating more computational power is at work, the mining difficulty rises to ensure that the rate of new block creation does not accelerate uncontrollably. Conversely, if the hash rate falls, the difficulty decreases to maintain a steady rate of block production.

The increasing hash rate, often a result of more participants joining the mining network or advancements in mining technology, directly impacts the mining difficulty and the probability of success for individual miners. As more miners compete to solve the puzzles, the likelihood of any single miner succeeding in finding a new block and receiving the block reward decreases. This higher competition often leads to miners pooling their resources in mining pools, enhancing their chances of success and achieving more consistent returns.

Simultaneously, this scenario presents a delicate balance between network security and the profitability of mining operations. A high hash rate generally indicates a secure network, as it suggests a broad base of miners contributing to the network's computational power, making it more resistant to potential attacks. However, the increased difficulty and competition can reduce profitability, especially for miners with higher operational costs. Miners are continually driven to seek more efficient mining methods, reduce operational expenses, and increase computational capabilities to stay profitable.

This balance is dynamic and closely tied to market conditions. In times of high profitability, such as when cryptocurrency prices are high, the network sees an influx of miners, driving up the hash rate and difficulty. The opposite happens during market downturns or when operational costs, like energy prices, increase, leading to a decrease in mining activity. This self-regulating mechanism of mining difficulty and hash rate serves to align individual mining incentives with the overall security and functionality of the network, illustrating a sophisticated design that ensures the long-term integrity and viability of these decentralized digital currencies. The intricacies of this relationship between mining difficulty and hash rate highlight the sophisticated and well-thought-out design of cryptocurrency networks, ensuring their stability and security over time.

The relationship between energy costs, mining efficiency, and profitability is a key consideration for miners. The cost of electricity can significantly vary across different regions, directly affecting the profitability of mining operations. Miners seek locations with cheaper electricity to reduce operational costs. Additionally, the efficiency of mining equipment plays a crucial role. More efficient hardware can process more calculations per unit of energy, reducing the overall energy requirement for mining. As technology advances, newer models of mining hardware become more energy-efficient, which can help reduce the total energy consumption of the network.

Environmental concerns arising from high energy consumption have led to a growing focus on the sustainability of mining activities. The carbon footprint of mining operations, primarily those reliant on fossil fuels, has raised questions about the environmental impact of cryptocurrencies. In response, there is an increasing shift towards renewable energy sources in mining. Utilizing renewable energy not only addresses environmental concerns but can also offer more stability and potentially lower long-term energy costs, benefiting both miners and the planet.

Events like block reward halvings and the evolving role of transaction fees have a significant impact on the long-term outlook of cryptocurrency mining. These factors play a pivotal role in influencing the economics of mining, its profitability, and energy consumption patterns.

Block reward halving is a feature embedded in the protocol of many cryptocurrencies, like Bitcoin, where the reward for mining a new block is halved at regular intervals. This mechanism is designed to control the rate of new currency creation, introducing a deflationary aspect to the cryptocurrency. From an economic standpoint, halving events have a profound impact. They effectively reduce the miner's reward for the same amount of mining effort, which can lead to a decrease in profitability, particularly for miners with higher operational costs.

The immediate aftermath of a halving event often brings uncertainty. Miners are faced with the decision to continue mining at lower profitability, upgrade to more efficient equipment, or cease operations. This period can lead to a temporary drop in the network's hash rate as less efficient miners exit the network. However, if the price of the cryptocurrency increases sufficiently post-halving, it can offset the reduced block reward, sustaining or even increasing the miners' profitability.

Projecting the future of mining profitability and energy consumption involves considering these halving events alongside the potential price movements of cryptocurrencies. As block rewards diminish over time, the reliance on transaction fees as a source of income for miners becomes increasingly crucial. This shift can alter the economic incentives for mining, potentially leading to changes in the network's energy consumption. If mining becomes less about new currency generation and more about transaction processing, the dynamics of energy use and the environmental impact of mining could shift.

The increasing importance of transaction fees over time is an essential aspect to consider. As the block rewards decrease, transaction fees are expected to represent a larger portion of the miner's income. This change could lead to a more stable and predictable source of revenue, as fees are based on network usage and transaction volumes rather than the fixed and diminishing schedule of block rewards. However, this shift also introduces new variables, as transaction fees can fluctuate based on network congestion and user behavior.

As we reach the conclusion of our exploration into the economics of cryptocurrency mining, it becomes clear that this field is not just a technological endeavor but a complex economic ecosystem. The key insights gained reveal a landscape where technological innovation, market dynamics, and energy concerns intersect, shaping the present and future of cryptocurrency mining.

Looking ahead, the future of cryptocurrency mining is poised to undergo significant changes. Technological advancements in mining equipment are expected to continue, potentially improving energy efficiency and reducing the environmental impact. The gradual reduction in block rewards will likely increase the relative importance of transaction fees, potentially leading to more transaction-focused mining activity. Additionally, the regulatory landscape is likely to evolve as governments and international bodies grapple with how to integrate cryptocurrencies into the global financial system while addressing environmental concerns.

The volatility of cryptocurrency prices remains a wildcard, capable of rapidly shifting the profitability landscape of mining. This volatility, alongside the decentralized nature of cryptocurrencies, ensures that the mining sector will continue to be an area of dynamic change and innovation.

In conclusion, the economics of cryptocurrency mining presents a fascinating blend of challenges and opportunities. It is a sector at the forefront of digital currency innovation, continually adapting to technological advancements, market forces, and environmental considerations. As cryptocurrencies themselves evolve and gain wider acceptance, the mining industry will undoubtedly play a pivotal role in shaping the future of digital finance.

Chapter 9
Bitcoin as a Medium of Exchange

Amidst the intricate weave of financial systems that define the modern era, Bitcoin stands out as an enigmatic yet transformative strand. Its birth heralded more than just an alternative asset; it offered a vision of a future unshackled from the tangible constraints of metal and paper. Bitcoin's cryptographic roots have planted the seeds for a landscape where value is not just exchanged but flows with ease in our digital lives.

The evolution of Bitcoin mirrors the evolution of money itself. It is a journey from the confines of a speculative investment to the liberating prospects of a medium for everyday transactions. Each step in Bitcoin's metamorphosis has been marked by growing pains and triumphs, reflecting its unique ability to adapt to the demands of a market in flux. The journey is emblematic of a broader shift in our collective understanding of what money is and what it can do. Bitcoin's transition from the speculative vaults of early adopters to the wallets of the global populace stands as a testament to this shift.

At the forefront of this endeavor are the intuitive digital wallets and exchanges that have begun to populate the market. For instance, platforms like Coinbase have garnered attention for their streamlined approach to buying, selling, and storing cryptocurrencies. By abstracting the intricacies of blockchain technology into a friendly and familiar interface, Coinbase has demystified Bitcoin for millions, offering a gateway through which novices can comfortably navigate the world of digital currencies.

Furthermore, the integration of Bitcoin into payment systems has been pivotal in cultivating its role as a transactional currency. Companies like BitPay and Square have led the charge, allowing merchants to accept Bitcoin payments without immersing themselves in the technicalities of cryptocurrency. BitPay's service converts Bitcoin into the merchant's local currency in real-time, eliminating the volatility risk – a significant concern for business owners. This seamless process has encouraged merchants worldwide to open their doors to Bitcoin, expanding its presence in the global economy.

Another beacon of innovation is the development of the Lightning Network, a layer-two protocol that operates atop the Bitcoin blockchain to enable faster and cheaper transactions. By taking transactions off-chain and settling them in batches, the Lightning Network addresses two of the most significant barriers to Bitcoin's scalability – transaction times and fees. Startups like Lightning Labs have been instrumental in pushing this technology forward, crafting user-friendly interfaces that allow even those without a deep understanding of blockchain to benefit from instant, low-cost Bitcoin transactions.

Perhaps the most compelling case study is El Salvador's bold move to adopt Bitcoin as legal tender. This unprecedented step was facilitated by government-backed wallet services that aim to simplify the use of Bitcoin for the nation's citizens and vendors. Despite the controversies and challenges, this national experiment serves as a real-world laboratory for how Bitcoin can be integrated into every facet of economic life.

These case studies collectively underscore a vital truth: the success of Bitcoin as a medium of exchange is as much a matter of design and user experience as it is of technological innovation. By focusing on user-friendly transactions and platforms that bridge the gap between complex protocols and practical usability, Bitcoin continues to carve a path toward becoming a currency not just of the present but of the future.

Despite its inception as a digital fortress for value, guarded by the complex algorithms and keys of cryptography, Bitcoin is now paving its way into the mundane transactions of everyday life. Yet, the current state of Bitcoin for day-to-day use is a tapestry of both progress and setbacks. While some cafes and online platforms welcome it with open digital arms, for the average consumer, spending Bitcoin on routine purchases is not without its hurdles. The volatility of its value, the labyrinthine nature of wallet addresses, and the sometimes sluggish transaction times form a triad of challenges that users must navigate.

For the regular user, these challenges are more than mere inconveniences; they represent significant barriers to adoption. The fluctuating price of Bitcoin can turn the simple act of buying a coffee into a speculative venture, making it difficult for both customers and merchants to commit to the currency for everyday transactions. Moreover, the technicalities involved in executing transactions—ensuring the right amount of digital currency leaves one wallet and arrives at another—demand a level of savvy and trust in the system that the general public is still cultivating.

Despite these challenges, the opportunities for Bitcoin in retail and commerce are vast and vibrant. As contactless payments become the norm and the digitization of money marches on, Bitcoin stands ready to offer an alternative to traditional financial structures. For the forward-thinking retailer, accepting Bitcoin can tap into a growing demographic of tech-savvy consumers and signal a commitment to innovation. Moreover, the decentralized nature of Bitcoin may offer merchants lower transaction fees compared to traditional credit card payments, making it an economically attractive option.

As Bitcoin matures, the landscape of its payment ecosystem is witnessing a significant expansion. The once niche market of Bitcoin payment options is now burgeoning with a multitude of services and technologies. Third-party payment processors like BitPay and CoinGate have emerged, offering merchants easy-to-integrate solutions that convert Bitcoin transactions into local currency almost instantaneously, thereby mitigating the risk associated with its volatility. Meanwhile, payment gateways are increasingly offering Bitcoin as a payment option alongside credit cards and PayPal, illustrating a growing confidence in its stability and longevity as a currency.

This ecosystem's growth is palpable among merchants and retailers who, driven by the demands of a tech-savvy consumer base, are increasingly adopting Bitcoin. Small businesses, e-commerce websites, and even some large multinational corporations are starting to accept Bitcoin, not just as a marketing ploy but as a response to a genuine and growing consumer trend. They are drawn by the allure of lower transaction fees, the security of blockchain-based transactions, and the ability to tap into a global market unrestricted by currency conversions and banking regulations.

Perhaps the most telling sign of Bitcoin's maturation is its integration with existing financial systems. Traditional banks, once wary of digital currencies, have begun exploring ways to incorporate Bitcoin into their offerings. Bitcoin ATMs are sprouting up around the globe, making the purchase and sale of the digital currency as convenient as a routine bank withdrawal. Some countries are even experimenting with the idea of central bank digital currencies (CBDCs), a nod to the transformative potential that Bitcoin and its underlying technology hold for the future of money.

The evolution of Bitcoin is marked by continuous adaptation, and among its most significant advancements is the development of the Lightning Network. This protocol operates as a second layer atop the Bitcoin blockchain and aims to solve some of the most pressing issues facing Bitcoin's scalability. By enabling a system where transactions can occur off the main blockchain, the Lightning Network allows for a dramatic increase in transaction speed and volume, addressing one of the core challenges of Bitcoin's original design.

The mechanism of the Lightning Network is both elegant and practical. It works by establishing payment channels between users that need only initial and final settlement transactions on the Bitcoin blockchain. Within these channels, transactions are nearly instantaneous and can occur in large volumes without burdening the main network. This is akin to opening a tab at a bar and settling the bill at the end of the night rather than paying for each drink individually, significantly reducing transaction time and fees.

This innovation is not just theoretical but is being applied in real-world scenarios with growing success. For small, everyday transactions, which would be impractical on the main blockchain due to high fees and slow processing times, the Lightning Network has become a game changer. Users can buy coffee, pay for online content, or send money to friends almost instantly and with negligible fees, similar to using digital wallets like Apple Pay or Venmo but with the added benefits of Bitcoin's decentralized architecture.

Real-world applications of the Lightning Network are burgeoning, with businesses and online services integrating it to provide customers with faster, cheaper payment options. For instance, certain online retailers and gaming platforms have begun accepting payments via the Lightning Network, catering to a user base eager for the advantages of Bitcoin without its transactional bottlenecks. Moreover, grassroots adoption is being driven by community initiatives, Lightning Network 'torch' payment chains, and even Lightning-enabled art and charity projects, demonstrating its versatility and user engagement.

At the heart of the Bitcoin experience is the wallet, a digital analog to the physical carriers of our cash and cards, and an essential tool in the facilitation of Bitcoin transactions. A Bitcoin wallet functions as the custodian of the private keys necessary to sign Bitcoin transactions, effectively granting access to Bitcoin addresses and enabling the user to send or receive the cryptocurrency. In essence, these wallets serve as a personal interface to the Bitcoin blockchain, allowing individuals to manage their digital currency.

The evolution of wallet technology has been driven by a dual mandate: ensuring robust security while maintaining user convenience. Early wallets were rudimentary by today's standards, offering basic functionality with a steep learning curve and security risks. However, as Bitcoin's popularity grew, so did the sophistication of wallet solutions, integrating advanced security measures such as hierarchical deterministic (HD) structures, multi-signature requirements, and hardware wallet options that store a user's private keys offline, significantly mitigating the risk of online theft.

Simultaneously, developers have strived to enhance the user experience, streamlining wallet interfaces and making them more intuitive. Modern wallets offer features like QR code scanning, NFC compatibility, and biometric security checks, all designed to make Bitcoin transactions as effortless as possible. The balance between robust security and user convenience is a delicate dance, with wallet developers constantly innovating to provide both in an accessible package.

In the marketplace, a variety of Bitcoin wallets vie for users' favor, each with its own set of features and trade-offs. There are 'hot' wallets, which are connected to the internet and prized for their convenience, and 'cold' wallets, which remain offline and are favored for their security. Desktop wallets provide robustness, mobile wallets offer portability, and web wallets promise accessibility from any internet-connected device. Furthermore, hardware wallets, resembling USB drives, represent the gold standard of security for many users.

Scalability concerns have long shadowed Bitcoin's rise, with skeptics pointing to the inherent limitations of its original design. The blockchain, a ledger of all past transactions, grows continuously, and with it grows the demand on network resources. This presents a challenge: how to process an ever-increasing volume of transactions without succumbing to bottlenecks that could throttle Bitcoin's utility as a daily currency. Each transaction requires confirmation by the network's miners, a process that, by design, is time and energy-intensive. This deliberate pacing—a feature implemented to ensure security—has become a stumbling block for scalability.

To overcome these hurdles, a suite of technical improvements has been introduced. Segregated Witness (SegWit) was one of the first major upgrades implemented to increase the block capacity without altering its size. It did so by removing signature information and storing it outside the base transaction block. This effectively shrinks the size of transactions, allowing more to fit within a single block and improving the overall capacity of the network. The aforementioned Lightning Network is another leap forward, creating a second layer where transactions can occur off the main blockchain, significantly speeding up the process and reducing the burden on the network.

Looking to the future, the prospects for scaling Bitcoin are intrinsically linked to the evolution of its technology and the ingenuity of its community. Developers continue to propose and test various solutions, from block size increases to more radical changes like the implementation of sidechains and sharding. As these solutions mature, they promise to further expand Bitcoin's capacity, ensuring its viability as a currency for the masses.

> **Bitcoin sharding is a hypothetical concept that would involve dividing the Bitcoin blockchain into smaller, more manageable pieces called shards. This would theoretically allow the network to process more transactions per second and improve its overall scalability.**

Yet, it's not just technological advancements that will scale Bitcoin; it's also the adoption of these technologies by the network's stakeholders—miners, developers, and users. The decentralized nature of Bitcoin means that consensus is key, and the future of Bitcoin's scalability will depend as much on the community's willingness to embrace change as on the developers' ability to innovate. The path forward is complex, but the continued focus on scalability ensures that Bitcoin remains at the forefront of financial technology, adapting and growing in line with the demands of its users.

Transaction fees are a pivotal aspect of Bitcoin's infrastructure, serving as an incentive for miners to process transactions. However, these fees can be a double-edged sword when considering Bitcoin's viability as a medium of exchange. High fees can deter users from adopting Bitcoin for smaller, everyday transactions, confining its use to larger, less frequent transfers where the fees are a smaller proportion of the transaction value. The fluctuation of these fees, often correlated with network congestion, adds a layer of unpredictability to the cost of transacting in Bitcoin, posing a challenge to its adoption for everyday commerce.

In response to this challenge, several strategies have been developed to minimize transaction costs. Users can adjust their fees based on their transaction urgency, opting to pay higher fees for faster confirmations or lower fees if they are willing to wait longer. Wallets have also become smarter, equipped with algorithms that can estimate the lowest fee required for a transaction to be confirmed within a reasonable timeframe. Additionally, innovations such as the Lightning Network provide a mechanism for conducting transactions with negligible fees, enhancing Bitcoin's potential for microtransactions and casual use.

The interplay between fees, security, and speed is a delicate balance central to the design of Bitcoin. Fees cannot be eliminated entirely without compromising the security provided by the network's miners. Similarly, speeding up transactions excessively could lead to security vulnerabilities. The Bitcoin community continues to explore this trade-off, seeking a sweet spot where fees are low enough to encourage widespread use, but high enough to maintain the robust security and decentralized nature of the blockchain.

The ongoing development and adoption of scaling solutions indicate a commitment within the Bitcoin community to address these concerns. By forging a middle path that respects the needs of both casual and power users, Bitcoin is striving to solidify its position not just as a store of value but as a practical medium of exchange. The balancing act between fees, security, and speed is an evolving narrative in the story of Bitcoin, reflecting its dynamic nature and its potential to adapt to the diverse needs of its global user base.

The road to mainstream adoption of Bitcoin is riddled with barriers, not least of which are the psychological hurdles that must be overcome. For many, the mere idea of a currency that exists purely in the digital realm is a departure from the tangible security of paper money and metal coins. There's a deep-rooted skepticism towards what can't be seen or touched, and the volatility that characterizes Bitcoin's market value often reinforces this wariness. The concept of a decentralized currency, free from the control of any government or financial institution, further challenges traditional notions of money and can be a source of unease for potential users.

To dismantle these psychological barriers, educational initiatives play a crucial role. Knowledge is the antidote to fear and misunderstanding, and as such, there's a growing movement dedicated to demystifying Bitcoin for the general public. These efforts range from online courses and seminars to workshops and informational content produced by cryptocurrency advocates. The aim is to provide a clear understanding of how Bitcoin works, its benefits, and its potential risks, thereby fostering a more informed and receptive attitude towards its use.

Regulatory challenges and compliance also present significant hurdles to Bitcoin's widespread adoption. The absence of a centralized regulatory framework for cryptocurrencies leads to a patchy landscape of laws and guidelines that vary wildly from one jurisdiction to another. For Bitcoin to move into the mainstream, it needs a regulatory environment that both protects consumers and fosters innovation. This involves not only the creation of clear regulations but also a dialogue between regulators, technologists, and the Bitcoin community to ensure that the regulatory framework supports and understands the nuances of cryptocurrency.

Compliance, too, is a complex issue. Bitcoin's pseudonymous nature raises concerns about its use for illicit activities. The development of compliance standards and the integration of know-your-customer (KYC) and anti-money laundering (AML) requirements into Bitcoin transactions are necessary steps for its acceptance by traditional financial institutions. These measures aim to align Bitcoin with the regulatory standards applied to conventional financial operations, thus easing the concerns of both regulators and the public.

Overcoming these barriers is an ongoing process, requiring a concerted effort from all stakeholders in the Bitcoin ecosystem. As the psychological, educational, and regulatory challenges are addressed, the path clears for Bitcoin to move from the fringes of internet subcultures into the wallets and lives of the broader global population.

The future outlook for Bitcoin as a medium of exchange is both fascinating and complex, surrounded by a haze of predictions and possibilities. In the rapidly evolving cryptocurrency space, emerging trends point toward a more integrated approach to digital currencies. There's a palpable sense of anticipation for the next generation of blockchain technology, which promises to streamline existing processes and introduce new functionalities. Innovations such as smart contracts and decentralized finance (DeFi) platforms are expanding the capabilities of cryptocurrencies well beyond simple transactions, potentially embedding them into the very fabric of economic activities.

Predictions for Bitcoin's role in the future economy vary widely, ranging from it becoming a universal digital gold, a store of value more than a means of transaction, to it fulfilling the vision of a fully-fledged currency for daily use. The increasing interest from institutional investors and the adoption by payment platforms and financial services suggest a growing confidence in its longevity and potential for integration into the mainstream financial system. As central banks and governments explore and launch their own digital currencies, Bitcoin may either face stiff competition or, alternatively, find its place solidified as the decentralized counterbalance to state-backed digital money.

Reflecting on Bitcoin's journey from an abstract concept to a tangible asset, it's clear that its path has been unlike any traditional currency. From its enigmatic beginnings on cryptography forums to the front pages of global financial news, Bitcoin has challenged preconceived notions about what money is and what it could be. It has prompted a reexamination of financial systems and has forced a conversation about the nature of value transfer in the digital age.

Closing thoughts on Bitcoin's evolution acknowledge the journey is far from complete. The future holds a myriad of technical, regulatory, and societal challenges to be navigated. Yet, the progress made thus far suggests a trajectory that continues to trend upwards, towards greater acceptance and integration. Whether Bitcoin ultimately becomes an everyday currency or remains a digital asset akin to gold, its impact on the economy and on global financial practices is indelible. As we look ahead, Bitcoin's legacy will likely be seen as a catalyst for a broad rethinking of how we earn, spend, and save in an interconnected digital world.

As we draw conclusions on Bitcoin's potential as a medium of exchange, it's clear that its journey from a cryptographer's vision to a globally recognized currency has been nothing short of extraordinary. Bitcoin has demonstrated the potential to redefine the very architecture of money, offering a decentralized alternative that challenges traditional financial institutions. It has shown that it can act not only as a store of value but also, increasingly, as a viable means for conducting transactions, ranging from the purchase of everyday items to large-scale international trade.

Reflecting on the path forward for Bitcoin in everyday commerce, there's a cautious optimism. The road ahead is paved with both technological advancements and potential regulatory frameworks that could either streamline or stifle its growth. Bitcoin continues to push the boundaries, driving innovation in payment systems and prompting a reevaluation of what constitutes money in a digital age. Its underlying blockchain technology has already begun to transform other sectors beyond finance, signaling the untapped potential that lies within.

As Bitcoin matures, its acceptance and integration into daily commerce will likely hinge on a confluence of factors. The continued development of user-friendly platforms, further enhancements in security and scalability, and clearer regulatory guidelines will all play a part in determining Bitcoin's place in the future economy. For many, Bitcoin represents more than just an alternative currency; it embodies a movement toward greater financial autonomy and inclusion.

In conclusion, Bitcoin's legacy may ultimately be defined by its ability to empower individuals, reshape traditional financial services, and provide a new lens through which we view the exchange of value in a rapidly digitizing world. Its story is ongoing—a narrative still being written by its users, developers, and the ever-changing dynamics of the global economy.

Chapter 10
Bitcoin's Evolution and Impact

In its infancy, Bitcoin was a rebel with a cause, a response to the 2008 financial crisis and a repudiation of the opaque practices of banking institutions. Its cryptographic underpinnings promised a system where trust was placed not in fallible institutions but in the immutable laws of mathematics. As Bitcoin matured, those who peered beyond its surface volatility began to see the underpinnings of a viable medium of exchange. The narrative of Bitcoin has since been marked by a gradual but unmistakable evolution: from the digital gold of a niche community of technologists and libertarians to a currency that promises to redefine the very act of transaction.

This evolution is not merely theoretical but practical, as Bitcoin begins to underpin a growing number of real-world transactions. Each day, more vendors, both online and in physical storefronts, are opening their doors to Bitcoin, recognizing its potential not just as an asset but as a functional currency. The journey from investment to transactional currency is marked by innovations and adaptations, each step forward solidifying Bitcoin's role as a medium of exchange. As we delve deeper into this journey, we come to appreciate Bitcoin not just for the value it holds but for the freedom and fluidity it offers in the exchange of that value.

The ascent of Bitcoin, charting a course from the arcane depths of cryptographic forums to the forefront of a financial revolution, is a testament to the ingenuity of its underlying protocol. Yet, for this digital currency to transcend its origins and achieve widespread adoption, it must bridge the chasm that separates a complex protocol from the everyday user. It is within this context that the role of user interface becomes paramount; a conduit through which the intricacies of Bitcoin are distilled into accessible, user-friendly applications.

The metamorphosis of Bitcoin into a currency accessible to all is largely dependent on the alchemy of transforming its complex protocol into a seamless user experience. Developers and designers have embarked on a mission to cloak the machinery of Bitcoin in interfaces that resonate with the simplicity and intuitiveness found in traditional financial services. Wallets that simplify the process of sending and receiving Bitcoin, exchanges that demystify the act of buying digital currency, and payment platforms that allow merchants to effortlessly integrate Bitcoin payments are the fruits of these labors. This push towards user-friendly interactions with Bitcoin is crucial in ushering in an era where digital currency is as straightforward as using a credit card or online banking service.

Case studies of successful protocol to user integration abound, each highlighting the potential of Bitcoin to integrate with the everyday financial activities of individuals and businesses. Take, for instance, the emergence of mobile wallets that empower users in developing countries to engage in global commerce, or peer-to-peer platforms that facilitate direct transactions without the need for intermediaries, each case showcasing the democratizing power of Bitcoin. There are stories of merchants in politically unstable regions adopting Bitcoin to maintain their financial autonomy, and tales of consumers using Bitcoin to protect their savings from the vagaries of volatile local currencies.

The journey of Bitcoin from a protocol spoken of in technical jargon to a currency that is used with the tap of a screen is one of continual refinement and innovation. It is a narrative marked by the relentless pursuit of simplification, ensuring that the power of the Bitcoin network is harnessed not through complexity, but through clarity and ease of use. As we delve into the ways in which this gap between protocol and user has been bridged, we bear witness to the transformation of Bitcoin from an obscure digital token into a currency poised for global adoption - a transformation that hinges not on the sophistication of the technology, but on the simplicity of its use.

The current state of Bitcoin for everyday use is a patchwork of rapid advancements and significant challenges - a landscape where visionaries and pragmatists alike are grappling with the currency's practical applications. While there are now cafes where one can purchase a latte with Bitcoin and online retailers that accept it for a wide range of goods, the reality is that these instances are islands in a vast sea of traditional commerce.

The challenges faced by users in regular transactions are not insignificant. Volatility looms large; the very characteristic that makes Bitcoin attractive to investors can be a hindrance when pricing goods and services. Transaction fees and processing times also present hurdles, as the scalability issues inherent in Bitcoin's original design are felt most acutely in times of peak demand. Furthermore, the user experience for making and receiving payments with Bitcoin can still be intimidating for the uninitiated, with wallet addresses and private keys forming a barrier of complexity for everyday users.

Yet, amidst these challenges lie tremendous opportunities for Bitcoin in retail and commerce. The potential for a currency that offers borderless transactions, reduced transaction fees for large purchases, and security from fraud is immense. For merchants, the appeal of Bitcoin is clear: it opens up a global market of consumers and provides protection from chargebacks. For consumers, particularly those in countries with unstable currencies or restrictive capital controls, Bitcoin offers a stable alternative for saving and spending.

Innovation is at the heart of Bitcoin's march towards everyday use. Contactless payments, instant transactions through the Lightning Network, and simplified security measures are on the horizon, promising a future where Bitcoin is as user-friendly as traditional electronic payment systems. As we explore the present and future of Bitcoin in everyday transactions, we look to a horizon brimming with potential - a horizon where Bitcoin is not just a digital curiosity, but a staple of daily commerce, reshaping the way we think about buying and selling in a connected world.

The growth of the Bitcoin payment ecosystem is a testament to the currency's enduring allure and its increasing acceptance within the broader financial landscape. This ecosystem, once a fledgling network of tech-savvy enthusiasts, has burgeoned into a robust marketplace with a myriad of payment options. Innovations such as Bitcoin ATMs, mobile wallet apps, and payment processors that convert Bitcoin transactions into local currency in real-time have significantly expanded Bitcoin's usability. These tools and services have not only simplified the act of transacting with Bitcoin but have also expanded its reach, enabling users to employ Bitcoin for everything from online shopping to remittances to even purchasing real estate.

The adoption of Bitcoin by merchants and retailers marks a significant milestone in its journey towards mainstream acceptance. From small independent cafes to multinational corporations, businesses are increasingly recognizing the benefits of adding Bitcoin to their payment repertoire. These benefits include access to a growing market of Bitcoin users, lower transaction fees compared to traditional credit card payments, and the elimination of certain types of financial fraud. As a result, the emblem of Bitcoin acceptance is becoming a more common sight at checkout counters and on website payment portals, signaling a shift in merchant attitudes towards digital currency.

Integration with existing financial systems is the most crucial factor in the Bitcoin payment ecosystem's growth. By forging partnerships with banks, payment gateways, and even mobile payment services, Bitcoin is shedding its outsider status and weaving itself into the fabric of established financial operations. This integration facilitates a smoother exchange between Bitcoin and traditional currencies, making it easier for consumers to use Bitcoin and for businesses to accept it. The result is a closing of the gap between the old and the new, between the fiat currencies of the past and the digital currencies of the future.

As we look at the burgeoning growth of the Bitcoin payment ecosystem, it's evident that the currency's potential is being realized in tangible, practical ways. Bitcoin's once-theoretical advantages are manifesting in real-world applications, indicating not just growth but a maturation of the currency and its supporting infrastructure. The landscape of

Bitcoin payments is a vibrant one, marked by innovation, adoption, and an ever-closer integration with the financial systems that underpin global commerce.

In the quest to address one of Bitcoin's most pressing challenges—its capacity for handling a vast volume of transactions—the development of the Lightning Network stands as a beacon of innovation. This protocol operates as a second layer atop the Bitcoin blockchain, a network designed for lightning-fast transactions that can occur off the main blockchain, thus enhancing Bitcoin's scalability and potential as a day-to-day currency.

The essence of the Lightning Network is simple in its genius: it allows for the creation of a system of payment channels that can be established between parties wishing to transact. These channels, once opened, enable an almost unlimited number of transactions between those parties, which are settled instantly and with minimal fees. The transactions conducted within these channels are only settled on the Bitcoin blockchain when the channels are closed, significantly reducing the transaction burden on the main network.

This off-chain approach to Bitcoin transactions is akin to settling a tab at a bar: instead of paying for each drink individually—each transaction requiring verification and adding to the bartender's workload—you simply open a tab at the start of the night and settle it when you're done. In this way, the Lightning Network holds the promise of turning Bitcoin into a currency that can handle the high transaction volumes required for widespread adoption, all without sacrificing the decentralized principles that form the bedrock of the Bitcoin network.

Real-world applications of the Lightning Network are already beginning to surface, providing a glimpse into a future where Bitcoin transactions are as quick and routine as using a credit card. Small, everyday transactions, previously impractical on Bitcoin's main network due to fees and processing times, are now feasible and economical. This has a particularly profound impact in regions with underdeveloped financial infrastructure, where the Lightning Network can offer a fast, reliable, and accessible payment method.

Early user experiences with the Lightning Network have been a mix of excitement and a recognition of the protocol's nascent state - there are technical hurdles to overcome and a user interface that must be refined. However, the promise of what the Lightning Network aims to achieve—a scalable, efficient, decentralized payment system—continues to drive its adoption and development. As we delve into the intricacies of this protocol and its applications, we bear witness to an evolving Bitcoin, one that is increasingly capable of meeting the demands of a global economy.

The landscape of Bitcoin is incomplete without the fundamental component of its architecture: the wallet. Bitcoin wallets function as the gatekeepers of cryptocurrency, serving as both the safeguard for a user's digital assets and the conduit through which all transactions pass. They are the personal interfaces to the Bitcoin network, enabling users to send, receive, and manage their Bitcoin holdings. With private keys securely stored, a wallet is the digital equivalent of a bank account, though it is the user, not the institution, who exercises complete control over its security and use.

As with any intersection of finance and technology, the balance between security and convenience is paramount in the design of Bitcoin wallets. On one end of the spectrum are hardware wallets—physical devices that store private keys offline, immune to online hacking attempts, epitomizing the 'cold storage' approach. On the other end are mobile and web wallets, which prioritize ease of access and are thus more suited to everyday transactions, embodying the concept of 'hot wallets'. Each type of wallet comes with its own set of security features and potential vulnerabilities, from two-factor authentication to deterministic signatures, all designed to provide users with the confidence to manage their Bitcoin securely.

The array of Bitcoin wallets available today is vast and varied, each with its own unique features tailored to different types of users. Some prioritize anonymity, obfuscating transaction details to preserve privacy, while others emphasize user experience, with intuitive interfaces that make navigating the Bitcoin network straightforward even for novices. There are wallets designed for traders, who need rapid, reliable access to the market, and those crafted for long-term investors, for whom security is the paramount concern.

Comparing various Bitcoin wallets requires a close examination of their features: the robustness of their security measures, the flexibility of their transaction capabilities, and the breadth of their compatibility with different devices and operating systems. From the minimalist to the feature-rich, from the open-source purists to the fully serviced platforms, the choice of a wallet can significantly influence one's experience with Bitcoin.

As we explore the development of Bitcoin wallets, we not only chart the technical evolution of these essential tools but also trace the shifting priorities and preferences of the Bitcoin community. The wallet is more than just a piece of software or hardware; it is the personal interface to the digital economy, evolving continuously to meet the demands of an increasingly sophisticated user base seeking to navigate the world of cryptocurrency.

Navigating the digital currency landscape, one encounters the pivotal challenge of scalability, a concern central to Bitcoin's evolution and its capacity to support a burgeoning user base. The original design of Bitcoin's network, while revolutionary, was not without its limitations. The blockchain, a ledger of all transaction history, can process only a finite number of transactions per block. As Bitcoin's popularity soared, this constraint led to bottlenecks, where transactions would outpace the network's ability to record them promptly, leading to delays and increased fees.

Addressing these scalability concerns has been the focus of intense technical innovation. Developers and engineers within the Bitcoin community have proposed and implemented several enhancements to streamline the network's efficiency. Solutions such as the Segregated Witness (SegWit) protocol have made transactions more compact, allowing more to fit within a single block. The development and gradual adoption of the Lightning Network, a second-layer protocol, promises to further alleviate the strain by enabling off-chain transactions that only settle on the blockchain as final balances, thus reducing the transaction volume that needs to be processed by the main network.

The future prospects for scaling Bitcoin hinge on a combination of technical ingenuity and community consensus. Proposed improvements such as Schnorr signatures aim to aggregate multiple signatures into one, thus reducing the data needed for multi-signature transactions. Sidechains, another proposed solution, offer the potential for transactions to occur on separate, adjacent blockchains, which could then be reconciled with the main Bitcoin blockchain, enhancing the overall capacity of the network.

Yet, the road to a fully scalable Bitcoin is not solely a technical journey; it is also a path forged by the decisions of its community. Each proposed upgrade must navigate the decentralized governance model that is the hallmark of Bitcoin, requiring widespread agreement from a diverse body of stakeholders, from miners and developers to end-users. This consensus-driven approach ensures that Bitcoin remains true to its foundational principles, even as it evolves to meet the demands of the future.

As we explore the multifaceted approaches to scalability, we glimpse the horizon of Bitcoin's potential, a vision of the currency not just as a store of value but as a medium of exchange for millions, or even billions, of people. In this journey lies the heart of Bitcoin's promise: to democratize finance, to offer an alternative to traditional currency, and to empower users across the globe with fast, secure, and low-cost transactions.

Transaction fees in the Bitcoin network serve as a critical incentive for miners, who validate and add transactions to the blockchain. However, these fees can also impact Bitcoin's viability as a medium of exchange. During periods of high network demand, fees can escalate, making small transactions economically unviable. This volatility in fees has sparked a debate about Bitcoin's practicality for everyday use, where the predictability of costs is paramount for both consumers and merchants. High fees could deter the adoption of Bitcoin for day-to-day transactions, such as buying a coffee or paying a utility bill, relegating it to the realm of significant investments or cross-border transfers where alternatives are costlier and less efficient.

Strategies for minimizing transaction costs are therefore crucial to enhancing Bitcoin's usability. Users can adjust their fees based on network congestion, choosing a lower fee during less busy periods, although this may result in slower confirmation times. Wallets with dynamic fee estimation algorithms can help in this decision-making process. Additionally, the adoption of Segregated Witness (SegWit) and batching multiple transactions into a single larger transaction are among the technical strategies that can reduce the individual cost of transactions.

The balance between fees, security, and speed is a delicate one. Fees must be high enough to incentivize miners to secure the network effectively, yet low enough to keep Bitcoin competitive against other payment methods. Moreover, the time it takes for a transaction to be confirmed on the blockchain is a crucial factor in user experience. Solutions such as the Lightning Network propose a compromise, offering near-instantaneous transactions with negligible fees, all while maintaining the security guarantees of the Bitcoin network through smart contract technology.

The path to mainstream adoption of Bitcoin, lined with the promise of a financial renaissance, is not without its obstacles. Among these, psychological barriers loom large, as traditional notions of currency are deeply ingrained. For many, the abstract nature of a digital currency, intangible and complex, presents a significant departure from the physical money that has been a fixture of commerce for centuries. The idea of entrusting one's wealth to the ethereal realm of the internet, to a currency that is not backed by a tangible asset or a central authority, requires a paradigm shift—a reimagining of what money is and can be.

To navigate these psychological barriers, educational initiatives have emerged as a crucial tool in demystifying Bitcoin. From online courses and workshops to informational content disseminated by enthusiasts and organizations, efforts are being made to break down the complexities of blockchain technology and elucidate the principles of digital currency. By fostering a greater public understanding of how Bitcoin operates and the benefits it offers, these initiatives aim to cultivate a more informed and receptive user base, one that can see beyond the uncertainty and recognize the potential of Bitcoin as a secure, decentralized alternative to conventional financial systems.

Yet, understanding alone is not enough to clear the path for Bitcoin's widespread acceptance. Regulatory challenges and the need for compliance present formidable hurdles. Governments and financial institutions worldwide grapple with how to integrate Bitcoin within existing legal and economic frameworks. The volatility of Bitcoin, concerns about its use for illicit transactions, and the potential implications for monetary policy have led to a cautious and sometimes resistant regulatory stance. As regulatory bodies work to establish guidelines that protect consumers without stifling innovation, the Bitcoin community continues to evolve, seeking ways to meet compliance standards while maintaining the currency's foundational ethos of decentralization and user autonomy.

The journey to overcoming the barriers to Bitcoin's mainstream adoption is a collective undertaking that spans the psychological, educational, and regulatory spheres. It's a journey that calls for a reevaluation of entrenched beliefs about money, a commitment to fostering knowledge and understanding, and a dialogue between the innovators of the digital currency space and the gatekeepers of the established financial order. As these barriers are confronted and gradually dismantled, the vision of Bitcoin as a ubiquitous medium of exchange comes into clearer focus, promising a future where financial transactions are more accessible, efficient, and equitable.

As we stand at the confluence of technological innovation and an economic revolution, the future outlook for Bitcoin as a medium of exchange is both intriguing and multifaceted. The cryptocurrency space is witnessing emerging trends that signal a shift in how digital assets are perceived and used. Decentralized finance (DeFi) platforms are expanding the utility of cryptocurrencies, while non-fungible tokens (NFTs) are redefining digital ownership, both pointing to a future where blockchain technology underpins a variety of financial instruments beyond simple currency.

The allure of Bitcoin extends beyond its architecture of zeroes and ones; it taps into a deep-seated desire for a financial system that is transparent, equitable, and under the direct control of the individual. In a world where trust in Governments and financial institutions has been eroded by crises and scandals, Bitcoin offers a form of currency that returns power to the people. This psychological appeal is formidable, galvanizing a community of users who are not just passive participants but active proponents of Bitcoin's vision for the future.

Envisioning Bitcoin at the center of a new financial ecosystem, we see a digital currency that paints a picture of a world where the currency flows as freely as information, unencumbered by national borders, immune to the ebb and flow of geopolitical tides. In this new realm, Bitcoin transcends its role as a mere store of value. It becomes a dynamic medium of exchange, integral to the fabric of daily commerce, enabling transactions that are not only expedient and secure but also liberated from the intricate web of international relations that so often constrains the movement of fiat currency.

The undercurrents of psychological forces shaping this financial renaissance are complex and profound. They stem from a deep well of governmental apprehension—a world stage where nations eye each other with a blend of skepticism and strategic calculation. Precluding the use of any one countries controlled currency or CBDC. The advent of Bitcoin, and the economic empowerment it promises, triggers a domino effect as one country's adoption prompts another to follow suit, spurred by the fear of economic disadvantage in an era of digital transformation.

Amidst this international narrative, the individual's role is equally transformative. The longing for financial autonomy has found resonance in Bitcoin's promise, appealing to those who seek to reclaim sovereignty over their economic destinies. This personal drive, when amalgamated with the collective desire for an alternative to the prevailing financial order, creates a potent force for change.

The impetus for Bitcoin's adoption thus goes beyond the mere mechanics of market operation. It is rooted in a collective ethos, a shared vision of what Bitcoin symbolizes— the return of financial control to the individual and the community, rather than to distant institutions and states. It is this vision that fuels the drive toward a world where Bitcoin sits at the heart of a new financial ecosystem, not as a distant ideal, but as a tangible reality that redefines our relationship with money.

In conclusion, the narrative of Bitcoin is still being written, its chapters unfolding in real time. With each transaction, each new user, and each technological breakthrough, Bitcoin is steadily carving out its place not just in the annals of economic history, but in the day-to-day transactions that stitch together the tapestry of human commerce. As we stand at the cusp of this financial frontier, it is clear that Bitcoin's journey from concept to cash is about more than just currency—it's about the redefinition of exchange in an interconnected digital age.

Section IV
Emerging Applications
and Challenges

Chapter 11
Decentralization and Trustlessness in Action

At the heart of Bitcoin lies a radical blueprint for reshaping the very foundation of economic exchange. This blueprint, drawn from the core principles of decentralization and trustlessness, proposes a system where transactions are no longer tethered to central authorities but are instead validated by a sprawling network of peers. It is a vision that challenges the monolithic structures of traditional finance with a model that disperses power across a distributed ledger, immutable and impervious to any single point of failure.

In today's economy, these principles take on a heightened significance. Decentralization empowers individuals to become their own custodians of wealth, sidestepping the need for intermediaries who have historically held sway over the mechanics of trade and trust. Trustlessness, a concept that might once have seemed counterintuitive in a world built on the bedrock of mutual confidence, now underpins a system where trust is encoded in the very fabric of transactions. In this new order, cryptographic proof replaces third-party verification, ensuring the integrity of each exchange and the inviolability of each participant's holdings.

As we reflect on these foundational tenets, the implications for our current economic landscape are profound. They herald a shift from centralized systems vulnerable to corruption and collapse to a decentralized paradigm where security and transparency are innate. This reimagination of economic principles, born from the digital crucible of Bitcoin, stands not only as a testament to human ingenuity but also as a challenge to the status quo - a challenge that beckons us toward a future where the control of money and the power it wields are returned to the collective hands of its rightful owners, the people.

In the burgeoning realm of digital finance, the principle of decentralization has transitioned from a theoretical ideal to a practical reality, manifesting in myriad ways that empower users and redefine traditional economic interactions. Peer-to-peer markets have flourished under this new paradigm, carving out spaces where individuals can engage in direct exchange without the oversight of central institutions. Such markets have seen exponential growth, as they cater to an increasing demand for transparency and autonomy, allowing users to trade everything from digital assets to real-world goods and services with unprecedented ease and security.

The transformative impact of these decentralized marketplaces is most palpable in regions where traditional fiat currencies waver under the weight of economic mismanagement. In countries beset by inflation or currency controls, Bitcoin and its peers offer a stable alternative, a haven for savings and a conduit for transactions that bypass volatile local currencies. Stories abound of individuals in these countries leveraging cryptocurrency to preserve the value of their labor and to engage in commerce on a scale that was previously unattainable, highlighting the tangible benefits that decentralization brings to areas long plagued by financial instability.

This decentralization has also given rise to novel business models that challenge the corporate structures that have dominated for centuries. Decentralized exchanges

(DEXs) now allow for the trading of digital assets without the need for a central authority, operating instead on smart contracts that execute trades automatically and immutably. These platforms have not only lowered barriers to entry for new participants but have also significantly reduced the risks of security breaches that centralized exchanges face.

Moreover, Decentralized Autonomous Organizations (DAOs) have emerged as the embodiment of organizational decentralization. These entities run on blockchain technology, governed by consensus mechanisms rather than by a board of directors, democratizing decision-making and ensuring that every stakeholder has a voice. DAOs are more than just a new type of business; they represent a rethinking of the very concept of the corporation, one that is less about hierarchy and more about collective action and shared governance.

Through these case studies – the rise of peer-to-peer markets, the stabilizing force of cryptocurrency in unstable economies, the emergence of DEXs, and the pioneering spirit of DAOs – we see the principle of decentralization not as an abstract concept, but as a living, breathing revolution in the way economic power is distributed and exercised. It is a shift that is not only technical but fundamentally human, reshaping not only markets and money but also the very fabric of societal and economic structures.

In Bitcoin's march towards financial ubiquity is the realization of a concept as old as commerce itself — trust. Or rather, the obviation of the need for it. This is the essence of 'trustlessness', a state in which the exchange of value no longer requires the assurance of a third party. It is here that cryptography, the bedrock upon which Bitcoin is built, shines; providing a framework where transactions are secure not because of mutual trust, but due to the mathematical certitude that underpins them. This cryptographic foundation ensures that a transaction, once recorded on the blockchain, is immutable — unchangeable, irrefutable, and entirely self-sufficient.

The implications of this shift are profound, particularly when examining the cumbersome process of traditional financial transactions. Consider remittances, a financial lifeline for families across borders. Traditionally, these cross-border transfers have been laden with high fees and long waiting times, routed through a labyrinth of banks and clearinghouses. Trustlessness cuts through these constraints, allowing remittances to flow directly between individuals, swiftly and with minimal cost. It dismantles the financial barriers, empowering migrant workers to send earnings home without the exorbitant toll taken by intermediaries.

Furthermore, smart contracts — self-executing contracts with the terms of the agreement directly written into lines of code — have emerged as a hallmark of trustlessness. They automate and secure transactions in a manner that was once the exclusive domain of lawyers and notaries. Smart contracts are not merely computer protocols; they represent a fundamental reordering of financial and legal processes. They ensure that a transaction or a contractual agreement is carried out flawlessly, with the terms enforced not by the might of legal systems, but by the unyielding logic of the code itself.

From the perspective of traditional financial hurdles, the paradigm of trustlessness introduced by Bitcoin and its cryptographic foundations is not an incremental improvement but a radical overhaul. It reimagines the flow of value and the execution of agreements in a world where the inefficiencies and vulnerabilities of trust-based systems are no longer a necessary evil but a relic of the past. Through the lens of trustlessness, Bitcoin is not just a new currency; it is the harbinger of a new order of economic interaction, one where the security and certainty of transactions are absolute.

In the realm of financial exchange, Bitcoin has established itself as a formidable bulwark against the overreach of censorship, embodying the principle of financial sovereignty. The digital currency's inherent censorship resistance has become a beacon for those seeking an alternative to the watchful eye of governmental control. In nations where freedom of transaction is stifled, Bitcoin has provided an avenue for the uninterrupted flow of capital, serving as a vital channel for the preservation of wealth and the assertion of individual economic rights.

Instances where Bitcoin has successfully circumvented governmental control are telling. In oppressive regimes, where conventional financial systems are heavily monitored and subject to stringent controls, Bitcoin has enabled citizens to bypass restrictions, moving funds with a degree of anonymity and security previously unimaginable. It is not merely a currency but a tool of economic liberation, challenging the traditional narrative that governments hold ultimate dominion over currency transactions. In this capacity, Bitcoin extends beyond a mere financial instrument to become an agent of freedom of speech and transaction, embodying the ethos that economic expression is a fundamental human right.

The protective qualities of Bitcoin have been particularly pronounced in times of economic crisis. Amid hyperinflation, where national currencies plummet in value and savings are eroded almost instantly, Bitcoin has offered a haven. In countries like Venezuela and Zimbabwe, Bitcoin adoption has surged as it provides a stable store of value, immune to the hyperinflationary spirals that ravage fiat currencies. Furthermore, in scenarios where political turmoil or policy shifts lead to asset seizure or devaluation, Bitcoin has enabled individuals to safeguard their wealth. Its decentralized nature means that unlike bank deposits or physical assets, Bitcoin is not susceptible to confiscation or the capricious whims of a volatile government.

The narrative of Bitcoin as a tool for financial sovereignty is not without its complexities, but the currency's ability to provide security and stability in uncertain times is irrefutable. As a medium that resists censorship and protects against the vicissitudes of economic mismanagement, Bitcoin has carved out a role not only as a revolutionary technological achievement but also as a sentinel of financial self-determination.

In the shadow of economic turmoil, where traditional monetary systems falter and currencies collapse, Bitcoin has emerged as an unexpected pillar of stability. This digital currency, once seen as the preserve of the tech-savvy and the speculative investor, has proven to be a bulwark for those caught in the maelstrom of financial crises. With its decentralized nature and global reach, Bitcoin has become a vehicle for stability in places where stability is a rare commodity.

The real-world efficacy of Bitcoin as an economic safe haven is perhaps best illustrated by the stark contrast it forms with the chaos of hyperinflation. In Venezuela, as the bolivar plummeted, rendering the nation's currency nearly worthless, Bitcoin emerged as a critical means of preserving wealth. It became a lifeline, allowing Venezuelans not only to maintain the value of their savings but also to conduct transactions in a currency that was immune to the hyperinflation engulfing their traditional monetary system. For many, Bitcoin's value was not just in its price but in its capacity to function where the national currency could not — as a stable medium of exchange.

A similar narrative unfolded during the Greek financial crisis, when capital controls left many Greeks with limited access to their own funds and a deep mistrust of the banking system. Bitcoin, with its promise of sovereignty over one's wealth, offered an alternative. It allowed individuals to move beyond the reach of a banking system in turmoil, offering both a means to secure assets and a pathway to engage in commerce amid the capital controls that constrained the nation.

The relationship between economic crises and the rise of Bitcoin adoption is telling. As economies wobble under the weight of mismanagement and fiscal uncertainty, Bitcoin's appeal has often surged. Its adoption is not merely a reflection of its growing acceptance but also a response to the need for an economic system that is not at the mercy of any single entity's control. In the face of capital controls, whether self-imposed or externally enforced, Bitcoin has provided individuals with a choice, a way to opt-out of a failing system, and into a new paradigm of financial self-reliance.

Bitcoin's ascent during these tumultuous periods is more than an anomaly; it is a testament to the digital currency's foundational principles. It underscores the growing recognition that in times of economic crisis, Bitcoin can offer what many traditional currencies cannot — a stable, accessible, and secure means of preserving and exchanging value.

As Bitcoin continues to carve its niche as a formidable force in the global economy, its march toward a fully decentralized future encounters a complex array of barriers. These obstacles not only test the resilience of its network but also the tenacity of its community.

The technical barriers are intricate and multifaceted. Bitcoin's network, designed to be secure and tamper-proof, faces challenges such as scalability, which in turn can lead to increased transaction fees and slower processing times during periods of peak demand. Innovations like the Lightning Network have emerged as a solution, creating a second layer where transactions can occur off the main blockchain, thus alleviating congestion and maintaining low fees. Other technical solutions, such as sidechains and updated consensus algorithms, are continually being developed and implemented to fortify the network against these challenges.

However, the journey toward decentralization is not soley hindered by technical hurdles. Sociopolitical challenges loom large as governments and financial institutions grapple with the implications of a currency that operates outside their control. The response across the globe is varied: some nations embrace the technology, seeing it as a catalyst for innovation and economic growth, while others clamp down with strict regulations or outright bans, viewing it as a threat to their fiscal control or a conduit for illicit activities.

Each of these responses shapes the global landscape in which Bitcoin operates, carving out a patchwork of acceptance and resistance. Advocacy and education have become paramount in this environment, with proponents of Bitcoin striving to demonstrate its potential benefits to society and its ability to function within the bounds of the law. In parallel, diplomatic engagement with regulators and policymakers is crucial to forging a path forward that respects the decentralized ethos of Bitcoin while addressing legitimate societal concerns.

As Bitcoin navigates these technical and sociopolitical barriers, the strength of its network and its community is put to the test. The global response thus far has highlighted not only the challenges of such a revolutionary shift in financial paradigms but also the opportunities it presents. In overcoming these barriers, Bitcoin does not merely aim to establish itself as an alternative monetary system; it seeks to redefine the principles of economic freedom and participation in a connected world.

The journey of Bitcoin, from its genesis as a manifestation of libertarian ideals to its burgeoning mainstream acceptance, charts a remarkable evolution of its use cases. Initially embraced by those who championed the virtues of a decentralized and ungoverned financial system, Bitcoin was a political statement as much as it was a technological innovation. It appealed to a community that valued privacy, individual autonomy, and skepticism of centralized authority—ideals that are etched into the very code upon which Bitcoin was built.

However, the narrative of Bitcoin has since expanded, reaching audiences far beyond its original libertarian enclave. Mainstream acceptance has been a gradual process, marked by milestones such as the recognition of Bitcoin by major financial institutions, its adoption by forward-thinking businesses, and an increasing presence in popular culture. As understanding and trust in Bitcoin grow, so too does its potential user base. This broadening appeal has translated into a diversification of how Bitcoin is used, moving it from the fringes of economic models to a more central role in global finance.

With this shift, the functionality of Bitcoin has expanded well beyond simple transactions. Bitcoin's underlying blockchain technology has opened up new avenues for innovation. It has become a platform for complex financial instruments, a means to raise capital through initial coin offerings, and a system for creating decentralized applications. In some economies, Bitcoin is starting to be recognized not just as an investment or speculative asset, but as a legitimate form of payment, comparable to conventional currencies.

Bitcoin is also increasingly seen as a tool for social empowerment, providing financial services to the unbanked and serving as a shield against economic instability in countries facing hyperinflation or financial repression. Digital currency is facilitating new ways to monetize content online, protect intellectual property, and crowdfund projects. The evolution of Bitcoin's use cases is a testament to the versatility of its technology and the adaptability of its community. It reflects a maturation of the currency, as it finds its place not only in the portfolios of investors but in the daily lives of people around the world.

The future of decentralized finance (DeFi) represents a radical departure from traditional banking, with Bitcoin standing at the forefront of this shift. DeFi extends the ethos of Bitcoin – direct, peer-to-peer interactions without central intermediaries – into a broader suite of financial services. It envisions an ecosystem where traditional offerings like lending, borrowing, and investing are reimagined through the lens of blockchain technology, offering a level of accessibility, inclusivity, and transparency previously unattainable.

The intersection of Bitcoin with the broader DeFi applications is pivotal. As the original cryptocurrency, Bitcoin laid the groundwork for this financial renaissance, demonstrating the viability of decentralized digital assets. DeFi applications take this a step further by leveraging blockchain technology to create complex financial systems – from automated loans and savings programs to synthetic assets and decentralized exchanges. Bitcoin, with its robust network and widespread acceptance, is a natural candidate for integration into these systems. It serves as a reliable store of value and medium of exchange within DeFi platforms, even facilitating collateralization and yield generation.

> **Yield is the return on an investment that reflects the income and/or capital gains generated from that investment over a specific period.**

The potential for a decentralized financial ecosystem with Bitcoin at its core is immense. Such an ecosystem would not only democratize access to financial services but could also increase market efficiency and reduce systemic risk by removing layers of intermediation. In DeFi, individuals retain control over their assets via cryptographic keys, eliminating the need for trust in fallible institutions and reducing the chance of centralized points of failure.

Moreover, DeFi applications built around Bitcoin could further enhance its utility. Bitcoin could serve as the gateway for a multitude of financial activities, with smart contracts automating transactions and enforcing terms in a trustless manner. This expansion could lead to a more integrated financial system where Bitcoin's liquidity and market cap bolster the entire DeFi landscape.

However, the emergence of such a decentralized financial ecosystem is challenging. Issues of scalability, interoperability, regulatory compliance, and user experience remain significant hurdles. The DeFi space is still young, and while its growth has been impressive, it must evolve to ensure robustness, security, and mainstream appeal. As we peer into the future of DeFi, we envision a world where financial systems are as open and accessible as the internet today – a world where Bitcoin and DeFi converge to create a new paradigm of financial interaction, one where the power dynamics of money are rewritten, empowering the individual as never before.

The transformative power of Bitcoin and decentralized finance reaches into the heart of communities, particularly touching the lives of those historically excluded from the traditional banking system. For the unbanked and underbanked — those who have been operating in the shadows of financial infrastructures — the advent of Bitcoin has been nothing short of revolutionary. With only a smartphone and an internet connection, individuals in even the most remote or impoverished regions can now access a global financial network. They can store value securely, engage in commerce, and connect to the global economy without the need for a bank account, which for many has been a barrier too high to surmount due to stringent requirements or lack of access.

Beyond just providing financial services, the rise of decentralized applications (DApps) has laid the foundation for community-driven innovation and local economic growth. These applications, running on blockchain platforms, enable communities to create tailored solutions to their unique economic challenges. From peer-to-peer energy trading platforms that allow neighbors to sell excess solar power to each other, to agricultural DApps that connect smallholder farmers directly with buyers, bypassing costly middlemen, the potential is vast.

DApps can also serve as a vehicle for community governance, allowing groups to make decisions in a transparent and democratic manner. This could fundamentally alter how communities manage shared resources, make collective decisions, and distribute wealth. The trustless nature of these applications ensures that the rules are applied equally to all, fostering a sense of fairness and collaboration.

The impact on individuals and communities is profound: economic empowerment leads to greater autonomy, and with it, the ability to innovate and pursue opportunities that were previously out of reach. This empowerment can catalyze a cycle of positive change, where increased economic activity supports further development and investment in community projects.

In this new era, where financial inclusion is no longer a lofty goal but a tangible reality, Bitcoin and decentralized applications stand as testaments to the potential of technology to serve humanity. They offer a glimpse of a future where financial tools are not just means of profit, but instruments of empowerment, where the true value lies not in the currency itself, but in the opportunities it unlocks for individuals and communities across the globe.

Navigating the evolving regulatory landscape is a pivotal challenge for the future of Bitcoin and the broader ecosystem of decentralized finance. As Bitcoin becomes more entrenched in the global economy, the push for regulatory clarity intensifies. Governments and financial institutions worldwide are grappling with the task of fitting a decentralized paradigm into a regulatory framework built for a centralized world. The path forward requires a delicate balance: crafting regulations that protect consumers and the integrity of the financial system without stifling the innovation and empowerment that decentralization brings.

Compliance, often viewed as anathema to the ethos of decentralization, is not necessarily a death knell for the freedoms that Bitcoin espouses. Rather, it can be a gateway to wider acceptance and stability. Thoughtful regulation can provide legitimacy, attracting more users and institutions to adopt Bitcoin, and can offer protection against fraud and manipulation. However, the decentralized nature of Bitcoin poses unique challenges for regulators accustomed to dealing with centralized entities. The absence of a central authority in Bitcoin means that traditional regulatory approaches must be reimagined. This might involve new forms of digital identity verification that preserve user privacy, or innovative reporting tools that leverage the transparency of the blockchain.

The role of regulation in shaping the future of decentralized Bitcoin transactions is crucial. It is a double-edged sword: too heavy-handed, and it could quash the very innovation that makes Bitcoin a transformative force; too lenient, and it could leave the system vulnerable to abuse. The ideal regulatory framework would recognize the unique attributes of Bitcoin — its openness, borderlessness, and neutrality — and would aim to safeguard these while mitigating the risks associated with digital currencies.

In this regard, the dialogue between the Bitcoin community, innovators, regulators, and policymakers becomes essential. Collaborative efforts are required to develop regulatory standards that align with the decentralized nature of Bitcoin. By engaging with regulatory bodies and participating in the legislative process, the Bitcoin community can help shape policies that reflect the nuances of decentralized finance.

The path forward is uncharted and complex, but with concerted effort and cooperation, it is possible to navigate the regulatory landscape in a way that preserves the decentralization and trustlessness at the heart of Bitcoin. Such a path would not only secure Bitcoin's place in the financial ecosystem of tomorrow but would also ensure that the revolutionary potential of Bitcoin — to democratize finance and empower individuals — is fully realized.

As we draw the curtain on the discourse of decentralization and trustlessness, we find ourselves reflecting on a narrative that is as much about reimagining the future as it is about the technology that underpins it. The advent of Bitcoin has catalyzed a seismic shift in the way we conceive of and interact with the very idea of value. Its impact is profound, with decentralization offering a redistribution of power away from central authorities to the periphery, to the individual user. Trustlessness has redefined the concept of security in transactions, negating the need for intermediaries and fostering a system where trust is embedded in cryptography and consensus algorithms rather than in fallible human institutions.

Envisioning a future shaped by these principles, we see a world where financial autonomy is the norm, and the barriers that once segregated individuals from the realm of global finance are dismantled. It is a future where the unbanked are no longer on the periphery but are active participants in a financial system that is inclusive, equitable, and free from the caprices of centralized control. Here, innovation thrives as developers and entrepreneurs build upon the open-source foundation laid down by Bitcoin to create services and applications that serve the common good.

As we stand at the threshold of a new economic epoch, the reflections on the decentralized economy that Bitcoin has pioneered compel us to consider our roles within it. The potential for a more equitable financial future is palpable, yet it requires more than mere passive observation; it calls for active participation. Each individual has the opportunity to engage with this evolving landscape, to become not only a user but a contributor to the ecosystem. Whether through mining, developing, investing, or simply using Bitcoin for daily transactions, participation is the fuel that drives the engine of decentralization.

The importance of education in this space cannot be overstressed. Understanding the intricacies of blockchain technology, the principles of cryptocurrency, and the implications of a decentralized economy is foundational to meaningful participation. Knowledge empowers users to make informed decisions, advocate for the technology, and contribute to its evolution in a way that aligns with its founding principles. Moreover, education acts as a bulwark against the misconceptions and misinformation that can cloud public perception of Bitcoin.

Advocacy, too, plays a critical role. It is through the voices of its community that Bitcoin can navigate the complex web of regulatory and societal challenges it faces. Advocates can influence policy, shape public opinion, and ensure that the decentralized economy remains true to its roots as it grows. They are the mediators between the world of Bitcoin and the broader society, translating the technical language of blockchain into the vocabulary of the everyday.

In the final analysis, the role of each individual in this economic shift is both significant and unique. The decentralized economy is not a monolith but a mosaic, composed of the diverse contributions of its participants. Whether one's role is that of a developer crafting the next generation of decentralized applications, a merchant accepting cryptocurrency, or an educator demystifying the concepts for newcomers, every action weaves into the larger narrative of Bitcoin. As we reflect on the transformative journey of Bitcoin, let us not be mere spectators to this revolution, but active participants, ready to leave our mark on the economic canvas of tomorrow.

Chapter 12
Challenges and Criticisms

In the grand tapestry of Bitcoin's history, its growth has been as remarkable as it has been fraught with challenges. From the nascent days of its inception to the towering presence it commands today, Bitcoin has weathered storms of skepticism and regulatory scrutiny. Yet, it stands resilient, a testament to the enduring vision of a decentralized financial future. This journey, marked by both milestones and obstacles, underscores the necessity not only to celebrate the strides made but also to confront the criticisms head-on.

The criticisms that Bitcoin faces are as diverse as they are complex, ranging from concerns over environmental impact to the intricacies of regulatory compliance. Addressing these is not a matter of defense but of necessity for the maturation of Bitcoin. It is through the crucible of critique that the strength of Bitcoin is tested and its framework fortified. Each challenge confronted and each issue resolved paves the way for a more robust and sophisticated system. Indeed, the very future of Bitcoin — its acceptance, integration, and evolution — is inextricably linked to how it responds to these critiques.

In recognizing the importance of addressing criticisms, the Bitcoin community and its developers are not merely engaging in a process of refinement; they are engaging in a dialogue about the future of finance itself. As Bitcoin continues to carve out its place in the global economy, the lessons learned from its challenges will shape not just its own trajectory but the contours of the emerging digital economy. It is a journey that demands rigor, creativity, and an unwavering commitment to the principles upon which Bitcoin was built.

As Bitcoin matures and solidifies its position in the financial world, an intriguing phenomenon surfaces, unique to its digital realm: the shrinking money supply. An inescapable reality of the Bitcoin ecosystem is the issue of lost wallets and irretrievable Bitcoins. Due to forgotten passwords, misplaced storage devices, or even the untimely demise of holders without a succession plan, a portion of Bitcoin's finite supply becomes inaccessible, effectively erasing a swath of the money supply from active circulation, a digital echo of sunken treasure lying inaccessible at the bottom of the sea.

The implications of this contraction are manifold. On the one hand, the scarcity may drive valuation, with each Bitcoin becoming more precious as the available supply diminishes. On the other hand, there are concerns about liquidity in the market. As coins become lost in the digital ether, the fluidity of the market could be impacted. This could lead to increased volatility, with larger price swings in response to market movements, and potentially make it more challenging for large-scale transactions to occur without influencing market price.

In response to the dwindling supply, various strategies have been proposed and implemented. Some in the community advocate for the continuous education of Bitcoin holders, emphasizing the importance of rigorous key management and the use of inheritable Bitcoin vaults. Others look to technological solutions, such as the development of more user-friendly wallets that minimize the risk of loss through advanced yet intuitive security features. Ultimately, the slow bleed of bitcoins into lost wallets could eventually spell the end of bitcoin as a currency in some far distant future.

The ongoing evolution of Bitcoin's protocol also offers potential remedies. Discussions within the community have considered the possibility of protocol changes that could, in theory, allow for the eventual recovery or replacement of lost coins. However, such drastic alterations come with contentious debate, as they touch upon the sacrosanct principle of the 21 million coin cap, a feature that many see as foundational to Bitcoin's value proposition.

Ultimately, the approach to Bitcoin's shrinking supply is characterized by a mix of education, innovation, and careful consideration of the currency's core tenets. As Bitcoin continues to navigate the uncharted waters of a digital economy, the community's response to these challenges will not only impact Bitcoin's liquidity and valuation but also define its role as a reliable store of value in the years to come.

In the midst of Bitcoin's ascent, the specter of quantum computing emerges, casting long shadows over the cryptographic bedrock that secures this digital currency. The realm of quantum computing, with its potential to perform calculations at speeds unfathomable to classical machines, presents a formidable challenge. The very algorithms that ensure the security of the blockchain and the immutability of Bitcoin transactions could, theoretically, be unraveled by the brute force of a sufficiently advanced quantum computer.

The cryptographic algorithms that Bitcoin employs— while currently robust against the onslaught of conventional hacking attempts—bear vulnerabilities when viewed through the lens of quantum capabilities. In particular, the algorithms responsible for Bitcoin's public-private key encryption could be susceptible to quantum attacks, which would enable the derivation of private keys from public addresses, threatening the sanctity of Bitcoin's security model.

This looming threat has not gone unheeded, and the Bitcoin community, ever vigilant, has rallied in response. Researchers and cryptographers are already deep into the fray, exploring the development of quantum-resistant solutions. These efforts range from post-quantum cryptographic algorithms, which could replace or augment Bitcoin's current encryption methods, to more comprehensive protocol upgrades that could fortify the network against quantum decryption.

The transition to quantum-resistant cryptography is a complex endeavor, layered with technical and communal challenges. Any proposed changes must maintain the integrity and trustlessness of the network while seamlessly integrating with existing infrastructure. The balance between evolution and preservation is delicate, with developers working meticulously to ensure that any amendments are as invisible as they are impenetrable.

The dialogue surrounding quantum computing and Bitcoin is not only a testament to the cryptocurrency's adaptability but also to its community's foresight. The push for quantum-resistant solutions is a proactive measure, one that seeks to safeguard Bitcoin's future before the threat materializes. In this, Bitcoin exhibits a hallmark of enduring technologies: the capacity not only to withstand change but to anticipate and evolve in the face of it.

The phenomenon of forking within Bitcoin's history is a tale of innovation, ideological divides, and the quest for consensus in a decentralized world. Hard forks, a term that evokes the image of a path diverging into two separate ways, represent significant changes to a blockchain's protocol that are not backward compatible. These forks can lead to the creation of a new blockchain that coexists with the old, each following its own distinct rules and vision for the future.

Hard forks are watershed moments that often stem from deep-rooted disagreements within the community over technical, philosophical, or strategic directions. The consequences are far-reaching: they can lead to confusion among users, potential security risks during the transition, and the dilution of resources and community focus between the old and new networks. Yet, they also embody the democratic nature of Bitcoin, allowing groups to vote with their computational power and choose the version of Bitcoin that aligns with their beliefs.

The history of Bitcoin forks is rich with such pivotal moments. From Bitcoin Cash to Bitcoin SV, these events are more than technical divergences; they are the manifestation of a community's struggle to find common ground. Each fork carries with it the collective hopes, aspirations, and dissent of a diverse group of individuals, all invested in the future of what money might be.

Central to the forking dilemma is the debate over consensus and the notion of unanimous agreement in protocol changes. In the ideal vision of Bitcoin, the community would move as one, with changes being universally agreed upon. However, the reality is more complex. Consensus in the Bitcoin network is not achieved through simple majority or unanimous agreement but through a more nuanced process that involves miners, developers, users, and businesses. This process is the crucible in which the future of Bitcoin is forged, a test of the strength and adaptability of its decentralized governance.

The debates that surround forks and consensus are emblematic of a broader conversation about governance in the digital age. They raise questions about how decisions should be made in a system that belongs to no one and everyone at the same time. As Bitcoin continues to evolve, the forking dilemma remains a central theme, challenging the community to navigate the delicate balance between unity and diversity, between the relentless march forward and the need to maintain a coherent, secure, and trusted financial network.

The energy consumption of Bitcoin mining has sparked a maelstrom of controversy, with critics pointing to its significant electrical appetite as a glaring contradiction to the digital age's push towards sustainability. The decentralized nature of Bitcoin requires a vast network of miners, each competing to solve complex cryptographic puzzles in a process that not only secures the blockchain but also consumes vast amounts of power. This energy-intensive procedure has drawn the ire of environmentalists and given rise to a heated debate about the place of cryptocurrencies in a world facing urgent climate challenges.

Detractors argue that the electricity used to mine Bitcoin often comes from sources that contribute to carbon emissions, exacerbating the environmental crisis. The criticism is not unfounded; certain mining operations do rely on coal-fired power plants, leaving a substantial carbon footprint. However, advocates for Bitcoin argue that this view does not take into account the complete picture. When compared to the traditional banking system with its extensive physical infrastructure, or the gold mining industry with its environmental degradation, Bitcoin's energy usage can be seen in a different light. The global banking system operates vast networks of branches, ATMs, and servers, all of which require energy — often more than what Bitcoin mining consumes.

The dialogue around Bitcoin's energy consumption is evolving, much like the cryptocurrency itself. Initiatives for renewable energy in mining are gaining momentum, with an increasing number of operations turning to solar, wind, and hydroelectric power to fuel their mining rigs. This not only mitigates the environmental impact but, in some cases, also makes economic sense as renewable energy costs decrease.

The controversy over Bitcoin's energy consumption continues, but it is clear that the future of mining is inexorably tied to the future of energy production. As the world pivots towards more sustainable energy sources, so too must Bitcoin if it is to maintain its place in the economy of tomorrow. The challenge is not insurmountable, but it requires a concerted effort from all corners of the Bitcoin ecosystem to ensure that the world's first decentralized currency can be both secure and sustainable.

Amidst the cacophony of criticisms leveled at Bitcoin, a chorus of counterarguments rises, addressing common misconceptions and illuminating the cryptocurrency's strengths. Critics often cite scalability and security as Achilles' heels, yet these concerns may not fully account for the ongoing advancements within Bitcoin's network. Scalability issues, once seen as an insurmountable barrier to Bitcoin's widespread adoption, are being actively addressed through solutions such as the Lightning Network, which promises to facilitate a higher volume of transactions with minimal fees. As for security, the robustness of Bitcoin's underlying blockchain technology has been proven over time, with the network withstanding a decade of scrutiny without succumbing to a single fatal flaw.

The role of education in this dynamic is pivotal. Misunderstandings about Bitcoin often stem from a lack of knowledge about how the technology operates. Educational initiatives play a crucial role in shifting public perception, demystifying the complexities of blockchain technology, and presenting Bitcoin in a more accurate light. By illuminating the mechanics of transactions, the mining process, and the principles of decentralized finance, education can dispel fears and foster a more informed and nuanced discourse around the cryptocurrency.

Moreover, highlighting Bitcoin's advantages over traditional financial systems is central to the counter-narrative. Bitcoin offers unparalleled financial sovereignty, with users having full control over their assets without the need for intermediaries. It presents a level of transparency that traditional banking cannot match, where transactions are recorded on an immutable public ledger. Additionally, Bitcoin provides an avenue for financial inclusion, reaching unbanked populations who have been left at the margins of the traditional financial ecosystem.

Bitcoin's potential to revolutionize financial systems around the globe rests not only on its technological merits but also on its ability to communicate and educate. By confronting criticisms with fact-based rebuttals, fostering an understanding of its technology, and drawing attention to its benefits, Bitcoin can continue to forge a path toward a future where it stands as a viable, secure, and scalable alternative to conventional currencies.

The ecosystem surrounding Bitcoin is one marked by an unyielding pace of technological innovation, addressing criticisms with a blend of ingenuity and foresight. Developers and technologists within the community are ceaselessly crafting solutions, be it through protocol upgrades that enhance scalability and efficiency or through the integration of new cryptographic techniques that bolster security against the looming shadow of quantum computing. These technological innovations are not mere patches; they are transformative changes that ensure Bitcoin's relevance in an evolving digital landscape.

Community-led initiatives are the lifeblood of Bitcoin's ecosystem. From grassroots advocacy to global educational campaigns, the community's efforts extend beyond the digital realm, fostering a culture of knowledge-sharing and open discourse. It is this collective endeavor that has given rise to a host of platforms, forums, and development groups dedicated to enhancing Bitcoin's usability and accessibility. These initiatives also serve as incubators for new ideas, where the next generation of Bitcoin applications and services is born and nurtured.

Adaptive measures for future-proofing Bitcoin are embedded in its DNA. The very structure of Bitcoin's governance — which requires consensus for significant changes — necessitates a proactive approach to potential threats and opportunities. The network's ability to adapt to an ever-changing environment is a testament to the strength of its decentralized model. This adaptability is reflected in the ongoing debates about the size of blocks, the implementation of second-layer solutions, and the meticulous research into post-quantum cryptography.

The future of Bitcoin, as these solutions and adaptations show, is not static. It is a dynamic interplay of technology, community, and a shared vision of an alternative financial system. With each challenge met and each adaptation adopted, Bitcoin is not just enduring; it is evolving, continually reasserting its position at the vanguard of the digital currency revolution.

The narrative of Bitcoin is punctuated with compelling case studies where communities and nations have faced and surmounted the currency's challenges. These stories paint a vivid picture of adversity, innovation, and triumph that lend credence to Bitcoin's potential.

One notable example is El Salvador, which took the unprecedented step of adopting Bitcoin as legal tender. This bold move was not without its challenges, including pushback from international financial institutions and internal debates about the implications for the nation's economy. However, by addressing these challenges head-on with clear regulatory frameworks and national programs to educate citizens about Bitcoin, El Salvador has positioned itself as a pioneer, exploring the benefits of cryptocurrency integration at a national level.

In addition to nation-states, local communities have also leveraged Bitcoin to address economic challenges. In places hit hard by economic sanctions or hyperinflation, such as Venezuela, Bitcoin has become a tool for survival. It has allowed citizens to transact beyond the reach of restrictive government policies, maintain the value of their savings, and engage in international commerce. The lessons learned here emphasize the importance of Bitcoin as an alternative financial system, particularly in areas where conventional economic structures have failed.

Another lesson from successful Bitcoin adoption comes from the way it has been integrated into everyday transactions. In cities like Zug, Switzerland, commonly referred to as "Crypto Valley," Bitcoin and other cryptocurrencies have been welcomed into the local economy, accepted as payment for goods and services, and even for municipal services. The success in Zug provides a blueprint for how a favorable regulatory environment, coupled with community engagement and education, can lead to widespread adoption.

The lessons from these diverse case studies are clear: understanding the specific needs and challenges of each community or nation is crucial. Moreover, the successful integration of Bitcoin hinges on the development of an ecosystem that includes supportive regulations, technological infrastructure, and educational initiatives. These factors work in concert to build trust in and adoption of Bitcoin, even in the face of adversity. Each case study serves not only as a testament to the resilience and flexibility of Bitcoin but also as a guide for other entities seeking to harness the potential of this groundbreaking technology.

As the discourse on Bitcoin's myriad challenges draws to a close, we take a moment to reflect on the indispensable role that overcoming adversity plays in the maturation and growth of any groundbreaking technology. Each challenge that Bitcoin has faced—from scalability woes to the drumbeat of regulatory scrutiny—has served as a crucible, tempering its structure and refining its community's resolve. It's a journey marked not by unalloyed triumphs but by learned resilience, where each setback has provided a valuable lesson in the evolutionary arc of this digital currency.

Bitcoin's journey is a testament to its malleability in the face of criticism. Rather than being brittle and unyielding, it has proven itself capable of dynamic change, evolving in accordance with both the changing landscape of technology and the shifting sands of user needs and expectations. The continuous interplay between challenge and change is the lifeblood of Bitcoin's development, driving innovations that enhance its usability, security, and relevance.

Looking forward, there is a wellspring of optimism about the role that Bitcoin will continue to play, despite—and indeed because of—the obstacles it encounters. This optimism is not rooted in naivety but in a track record of resilience and innovation. From the implementation of the Lightning Network to the growing acceptance of Bitcoin in various sectors of the economy, there are tangible signs that Bitcoin is not just weathering the storm but thriving within it.

In the conclusion of this chapter, we acknowledge that the path forward for Bitcoin is not one of unchallenged ascendency but rather one that will be defined by its ability to navigate and surmount the obstacles that arise. It is through this ongoing process of overcoming and adapting that Bitcoin will continue to carve out its place in the financial ecosystem, not just as a novel asset but as a transformative force that redefines what a global currency can be.

As the discourse around Bitcoin reaches a moment of introspection, it becomes clear that critical analysis is not merely an academic exercise, but a vital component in the maturation of this digital currency. The scrutiny that Bitcoin endures is a powerful force, compelling it to constantly reassess and strengthen its protocols, community, and place within the broader financial landscape. This process of rigorous examination and debate ensures that Bitcoin does not stagnate but continues to evolve, meeting the needs and expectations of a diverse and growing user base.

The Bitcoin ecosystem is distinguished by its vibrant community of users, developers, thinkers, and entrepreneurs, all of whom play an active role in shaping its trajectory. Encouraging this community engagement in problem-solving is crucial. The decentralized nature of Bitcoin means that its future is written not by a select few, but by the many who invest their time and intellect into its continuous improvement. From addressing technical challenges to navigating the complexities of global regulations, the collective wisdom and creativity of the Bitcoin community are its most valuable assets.

Envisioning the future of Bitcoin, one can see a landscape where challenges are not roadblocks but catalysts for innovation. The issues faced today, and those that will emerge tomorrow, set the stage for breakthroughs that may redefine what is possible with blockchain technology. This future, brimming with potential, is built on the foundation of a currency that is not only a tool for financial transactions but also a platform for endless possibilities.

The story of Bitcoin is one of continual transformation, driven by the very challenges that test its strength. It is a narrative that underscores the remarkable potential of an open-source project to not just endure but to thrive as it is shaped by the hands of a global community dedicated to the vision of a decentralized and empowered future.

Chapter 13
Navigating the Diverse World of Cryptocurrencies

As we embark into the diverse world of cryptocurrencies, we will explore the most prominent and enduring digital currencies shaping our current financial landscape. As of the time of this publication, the cryptocurrency market has burgeoned into a vast and varied ecosystem featuring a multitude of coins and tokens, each with its own unique purpose and technological underpinnings. Our exploration delves into these cryptocurrencies, not just to understand their current functions but to grasp the broader implications they hold for the future of finance and economics.

As such, we will present a couple of coins that highlight each functional category. The categories chosen are a window into understanding some of the diverse functionalities and roles in the digital economy. From coins like Bitcoin, which have established themselves as pioneers in the field, to the tokens that are pushing the boundaries of blockchain technology, each category represents a different facet of the crypto world. However, it is crucial to acknowledge that these categories are not rigid or exhaustive. The landscape of cryptocurrency is in constant flux, with innovations continually emerging, blurring the lines between established categories and creating new ones.

This book does not merely list cryptocurrencies but seeks to classify them in a way that provides clarity and insight into their roles, risks, and potential. Yet, it's essential to recognize that this classification is based on the current state of the market and the understanding of these technologies at the time of writing. The world of cryptocurrencies is rapidly changing, so what may be true today might change tomorrow due to new developments.

As we delve deeper into the realm of cryptocurrencies, it becomes crucial to distinguish between two fundamental concepts: coins and tokens. While these terms are often used interchangeably in casual conversation, they represent distinct entities in the world of digital finance.

Coins are digital currencies that operate on their own dedicated blockchain. These blockchains serve as the foundational technology, recording all transactions and maintaining the security and integrity of the currency. Bitcoin and its blockchain bearing the same name is a prime example of a coin. It functions independently on its network, where transactions are verified and recorded. Similarly, other coins, like Ethereum and Litecoin, operate on their own distinct blockchains. They not only serve as mediums of exchange but also provide platforms for various blockchain-based applications, expanding their utility beyond simple financial transactions.

Tokens, on the other hand, are a different breed within the cryptocurrency ecosystem. Unlike coins, tokens are created on existing blockchains. The most common platform for token creation is Ethereum, which introduced the concept of smart contracts. These smart contracts allow for the creation of tokens that can represent a wide array of assets or functionalities. For instance, tokens can be designed to represent a stake in a project (security tokens), a right to use certain services within a platform (utility tokens), or even digital art and collectibles (non-fungible tokens, or NFTs).

The versatility of tokens means they can serve purposes that go far beyond mere currency transactions. They can embody ownership, rights, or access to services, and they can be tailored to the specific needs of a project or community. This flexibility has led to a proliferation of tokens, each designed with specific functions and purposes within the ecosystem in which they operate.

In essence, the critical difference lies in their underlying infrastructure and intended use—coins, with their independent blockchains, primarily function as digital money. Tokens, being built on these pre-existing blockchains, are more diverse in their roles and applications. This distinction is pivotal in understanding the broad spectrum of opportunities and functionalities that the world of cryptocurrencies offers. As we explore further, the diverse and innovative uses of both coins and tokens within this digital landscape become increasingly apparent, showcasing the vast potential of blockchain technology in reshaping our economic and social systems.

Payment Cryptocurrencies

Cryptocurrencies have emerged as assets that function as alternatives to currencies, with a major focus on transaction facilitation. They intend to simplify and modernize the value exchange process.

Bitcoin, which has previously been fully addressed in this book, is at the vanguard of this category. Bitcoin has the distinction of being the most well-known cryptocurrency in the world. It pioneered the concept of digital currency, paving the path for others to follow suit. Bitcoin's concept is one of delivering decentralized transactions free of the control of any central authority. Its function as a store of value and a means of exchange has influenced global understanding and adoption of cryptocurrencies.

Litecoin is another major contender in this field. Litecoin, sometimes referred to as the silver to Bitcoin's gold, provides significant advantages in terms of transaction efficiency. It was created to enhance the Bitcoin protocol by providing faster transaction confirmation times and a more efficient mining process. Litecoin does this through its unique hashing algorithm and a lower block generation time, which allows for faster transaction processing. As a result, Litecoin is an appealing option for those looking for speedier and more efficient transaction capabilities in their daily digital exchanges.

Bitcoin and Litecoin both demonstrate the core role of payment cryptocurrencies, which is to provide a digital and decentralized alternative to traditional fiat currencies. They allow users to deal across borders without the use of intermediaries, indicating a future in which digital currencies promote the flow of global commerce with increased efficiency and accessibility. Payment cryptocurrencies such as Bitcoin and Litecoin are key contributions to this paradigm shift in financial transactions as the globe continues to gravitate toward digital solutions.

Stablecoins

In the dynamic world of cryptocurrency, stablecoins occupy a unique niche, designed to marry the digital efficiency of cryptocurrencies with the stability of traditional fiat currencies. They achieve this by being pegged to stable assets like the US dollar or gold, thereby maintaining a consistent value. This pegging is crucial in mitigating the high volatility often associated with more mainstream cryptocurrencies such as Bitcoin and Ethereum, making stablecoins a more predictable digital asset for everyday transactions and financial operations.

Despite their advantages, stablecoins face significant challenges. A primary concern is the risk of de-pegging, where the value of the stablecoin could significantly deviate from the asset it's pegged to. Such a scenario could lead to instability and a crisis of confidence among its users. Furthermore, maintaining the peg requires some level of centralization, as a central authority must manage the reserves of the pegged asset. This centralization is a point of contention within the cryptocurrency community, which primarily values decentralization.

The maintenance of the peg in stablecoins is a delicate and complex process. It entails holding a reserve of the asset (or a basket of assets) to which a controlling organization has pegged the stablecoin at an equivalent or higher value. This reserve acts as a guarantee for the stablecoin's value. Therefore, regular audits and transparency reports are essential to maintaining trust among users and ensuring that the reserves are sufficient and appropriately managed. This transparency is crucial for maintaining user confidence. A failure of user confidence can lead to a destabilizing and destructive run on the coin, especially if it turns out that 100% redeemable reserves did not back it. This aspect is critical in the cryptocurrency context, where the traditional banking concept of fractional reserves is incompatible. The cryptocurrency community, valuing transparency and full reserve practices, would likely reject any attempt by a central bank-based stablecoin to introduce the flawed concept of fractional reserves into this space.

Tether (USDT) and USD Coin (USDC) are two of the most prominent examples of stablecoins. Tether, with its global operations, claims that each of its USDT tokens is backed by a corresponding US dollar in its reserves. However, this claim has been subject to scrutiny and controversy, especially regarding the transparency and verification of its reserves. On the other hand, USD Coin, operating primarily within the US, adheres to a stricter regulatory environment, aiming for higher transparency and regulatory compliance. This approach seeks to provide users with greater assurance regarding the stability and reliability of the coin.

In summary, stablecoins like USDT and USDC are vital in bridging the gap between traditional and digital finance. They offer stability in the often turbulent cryptocurrency market, making them appealing for both everyday transactions and as a safeguard against volatility. However, the challenges they face, particularly in maintaining their peg and ensuring operational transparency, are critical aspects that will determine their success and acceptance in the broader financial landscape.

Utility Tokens

In the diverse world of cryptocurrencies, Ethereum's ETH and Cardano's ADA occupy a unique position. While they are widely recognized as coins due to their native blockchains, they also function as utility tokens within their respective ecosystems. This dual role is a testament to the multifaceted nature of these digital assets. Unlike cryptocurrencies, which are designed primarily as mediums of exchange, utility tokens facilitate access to services and features within specific digital ecosystems. They act as "keys" to unlock functionalities in blockchain-based platforms and applications, diverging from the traditional role of currencies.

A prominent example is Ethereum (ETH), the lifeblood of the Ethereum blockchain. Renowned for hosting decentralized applications (dApps) and executing smart contracts, ETH is used as 'gas' for these operations. It is essential for powering the Ethereum network's activities, making its demand directly linked to the blockchain's usage. The escalating development of dApps and projects within the Ethereum ecosystem underscores ETH's critical role as a utility token.

Cardano (ADA) provides another illustration of the potential of utility tokens. Developed with a focus on peer-reviewed scientific research, Cardano aims to offer a more secure and sustainable platform for smart contracts and dApps. ADA, Cardano's native token, supports transactions and smart contract functionalities within its network. With its emphasis on scalability, interoperability, and sustainability, Cardano seeks to address some limitations and challenges faced by earlier blockchain platforms, including Ethereum.

What sets these platforms apart is their ability to host other utility tokens. Both Ethereum and Cardano allow for the creation of new tokens on their networks, which require ETH or ADA to function, particularly for transaction processing or 'gas' fees. This aspect of their ecosystems opens up a plethora of possibilities for decentralized applications and value exchanges.

The amount of control that their creators retain,

however, is a crucial component in the world of utility tokens. Typically, the development team or organization behind a utility token exercises significant authority over its functionality, governance, and distribution. This control ranges from defining the token's utility within the ecosystem, deciding the total supply and distribution mechanisms, and setting up governance rules. Despite the decentralized nature of blockchain, this introduces a centralized facet to utility tokens, where creators hold sway over the token's trajectory.

This level of control can be advantageous for efficient decision-making and adaptation of the token to the project's needs. Yet, it also raises concerns about centralization and the potential for misuse of power. Token holders often rely on the creators' credibility and decision-making, highlighting the importance of trust and transparency in these projects.

As the blockchain ecosystem continues to evolve, the significance of utility tokens is expected to rise, marking new avenues for digital engagement and value exchange. Their success and trustworthiness largely hinge on the balance between decentralized innovation and the centralized control exerted by their creators.

Security Tokens

Security tokens are a burgeoning cryptocurrency category that represents a digital version of traditional securities. These are investment vehicles that are frequently linked to an underlying asset or an ownership share in a project or firm. The primary attraction of security tokens is their combination of blockchain technology with the legal and regulatory framework of traditional financial assets. This connection aims to improve the transparency, efficiency, and accessibility of investing processes.

tZERO is one of the market's first and most significant companies in the security token space. Founded as an Overstock.com subsidiary, tZERO has been at the forefront of integrating blockchain technology into the field of regulated financial products. It provides a platform for trading security tokens, assuring regulatory compliance, and offering investors a more transparent and efficient trading experience. tZERO has been a leader in the security token arena, proving the practicality and potential of blockchain technology to modernize traditional financial markets.

Polymath (POLY) is another significant participant in the security token space, offering a platform intended exclusively for generating, issuing, and managing compliant security tokens. It simplifies the complicated legal and technical issues associated with tokenizing financial instruments. To facilitate the issuing of regulatory-compliant tokens, Polymath's ecosystem brings together issuers, legal delegates, smart contract developers, and investors. This ecosystem approach, combined with Polymath's emphasis on compliance and security, presents it as a critical enabler for enterprises wishing to use blockchain technology for regulated financial products.

Platforms like tZERO and Polymath, which represent security tokens, are establishing a new market niche in the cryptocurrency sector. They are bridging the gap between blockchain technology's revolutionary potential and the regulatory frameworks that control financial assets. This convergence promises to democratize access to investment opportunities, lower costs and restrictions for issuers and investors, and improve the securities market's overall integrity and efficiency. Security tokens are set to play an increasingly important part in the future of banking as the regulatory framework evolves with technological improvements.

Privacy Coins

Privacy coins represent a specialized segment within the broader cryptocurrency world, with a focused objective: ensuring anonymity and privacy in financial transactions. Unlike more transparent blockchain networks like Bitcoin or Ethereum, where transaction details and wallet addresses can be traced and analyzed, privacy coins offer enhanced privacy features to shield this information, making transactions untraceable and wallet balances private.

Monero (XMR) is a prime example of a privacy coin renowned for its robust privacy features. It employs sophisticated cryptographic techniques, like ring signatures and stealth addresses, to obscure transaction details. Ring signatures mix a user's account keys with public keys obtained from Monero's blockchain to create a unique group of signers, making it virtually impossible to link a transaction to a specific user. Stealth addresses, on the other hand, are one-time-use addresses created for each transaction on behalf of the recipient, ensuring the true destination of the transaction remains hidden. These features make Monero a favored choice for users seeking complete privacy in their financial dealings.

Zcash (ZEC) offers another approach to privacy, setting it apart from other privacy coins. While Zcash transactions can be transparent, similar to Bitcoin transactions, it also allows users to send funds privately using its shielded transaction feature. The network can validate transactions without disclosing sensitive information about the sender, recipient, or transaction amount thanks to a technology known as zk-SNARKs (Zero-Knowledge Succinct Non-Interactive Arguments of Knowledge). This dual feature of Zcash offers users the flexibility to choose between transparency and privacy, catering to a broader range of needs and preferences.

Privacy coins like Monero and Zcash play a critical role in the cryptocurrency ecosystem. They cater to a segment of users who prioritize privacy above all else, offering them alternatives that protect their financial privacy. As discussions around privacy and anonymity in the digital space continue to gain prominence, the relevance and importance of privacy coins are likely to grow, highlighting the diverse needs and preferences within the cryptocurrency community.

DeFi Tokens

DeFi tokens are at the forefront of a significant shift in the financial world, serving as the backbone of decentralized finance (DeFi) platforms that aim to replace or augment traditional financial systems with decentralized alternatives. These tokens are not just mediums of exchange but are integral to the functioning and governance of DeFi platforms. They often facilitate various financial services like lending, borrowing, and trading, all executed on blockchain technology without centralized intermediaries.

Uniswap (UNI) is a prominent example in the DeFi space, particularly in the realm of decentralized exchanges (DEXs). Uniswap's native token, UNI, plays a crucial role in its ecosystem. It's not only used for transactions and liquidity provision on the platform but also gives holders governance rights, allowing them to participate in decision-making processes concerning the development and changes within the Uniswap protocol. This level of direct involvement is a departure from traditional finance models and a testament to the ethos of decentralization. Uniswap's innovative automated liquidity protocol has set a standard in the DeFi space, offering users a seamless way to exchange cryptocurrencies without relying on traditional market-making mechanisms.

SushiSwap (SUSHI) represents another intriguing facet of the DeFi ecosystem. It began as a fork of Uniswap but quickly carved out its niche, primarily due to its community-driven approach. The SUSHI token not only facilitates transactions and liquidity on the platform but also entitles holders to a share of the platform's transaction fees, acting as a form of passive income. This model promotes a strong sense of community ownership and participation, as SUSHI holders effectively become stakeholders in the platform's success. SushiSwap's growth and evolution offer a case study on how community engagement and incentives can drive innovation and adoption in the DeFi sector.

DeFi tokens like UNI and SUSHI are reshaping the landscape of financial services, offering decentralized, transparent, and community-driven alternatives to traditional finance. As these platforms grow and evolve, they challenge the norms of financial operations, opening up possibilities for more inclusive and equitable financial systems. DeFi tokens are not just digital assets; they are the linchpins of a rapidly developing ecosystem that promises to redefine finance for the digital age.

Meme Coins

Meme coins, a unique and somewhat unexpected phenomenon in the cryptocurrency world, started as humorous takes on the more serious and technologically driven cryptocurrencies like Bitcoin and Ethereum. These coins may have originated as jokes or tributes to popular internet memes, but against many odds, they have catapulted to popularity and have become significant players in the crypto market.

Dogecoin (DOGE), perhaps the most well-known meme coin, epitomizes this journey from a lighthearted joke to a mainstream cryptocurrency. Initially created in 2013 as a playful nod to the then-popular 'Doge' meme featuring a Shiba Inu dog, Dogecoin was not intended to be taken seriously. However, its friendly branding and the welcoming community that formed around it quickly propelled it to unexpected heights. Numerous high-profile endorsements and social media buzz, particularly on sites like Twitter and Reddit, contributed to Dogecoin's rise to prominence. This surge in popularity not only increased its value but also solidified its position as a legitimate, though still somewhat whimsical, player in the crypto market.

Shiba Inu (SHIB), another meme coin that has gained notable attention, offers an exciting study of virality in the crypto world. Much like Dogecoin, Shiba Inu leverages the appeal of the Shiba Inu dog and has strong community support, often referred to as the "SHIBArmy." Its rise can be attributed to a combination of clever marketing, community-building efforts, and the broader appeal of meme culture within the crypto community. Shiba Inu's growth is indicative of how cryptocurrencies can transcend traditional financial logic, riding the waves of social media trends and community sentiment to achieve value and recognition.

The emergence and success of meme coins like Dogecoin and Shiba Inu highlight an intriguing aspect of digital currencies: their value and relevance can be as much about social phenomena and community sentiment as they are about technology and utility. While the same level of technological innovation may sometimes back these coins as their more serious counterparts, their ability to capture the imagination and enthusiasm of a broad user base makes them an undeniable force in the cryptocurrency landscape.

Layer 2 Solutions

Layer 2 solutions have emerged as vital protocols that enhance the scalability and efficiency of existing blockchains. These innovative solutions are pivotal in addressing the challenges of transaction speed and network congestion that often plague popular blockchains.

Polygon, previously known as Matic Network, stands as a sterling example of this innovation. It functions as a multi-chain scaling solution for Ethereum, focusing on improving its capabilities. Polygon's framework allows for the construction and interconnection of Ethereum-compatible blockchain networks. It's a transformative approach that extends Ethereum's functionality without compromising its underlying principles. By offering solutions like sidechains, Polygon effectively manages to reduce transaction costs and enhance speed, making Ethereum more accessible and usable for a broader range of applications.

Another significant player in this field is Optimism, which has positioned itself as a leading Layer 2 scaling solution for the Ethereum blockchain. Optimism harnesses the power of Optimistic Rollups, a technology designed to increase transaction throughput significantly. This innovation reduces the Ethereum network's congestion and the associated transaction fees. By executing transactions off the main Ethereum chain while still under its security umbrella, Optimism ensures a balance between efficiency and the decentralized security that Ethereum is known for. This approach not only enhances the user experience but also expands Ethereum's potential for a more comprehensive array of decentralized applications and use cases.

Both Polygon and Optimism represent the evolving nature of blockchain technology. Their contributions are instrumental in pushing the boundaries of what established blockchains like Ethereum can achieve. By solving critical issues of scalability and cost, these Layer 2 solutions are essential in driving the mass adoption of blockchain technology and its integration into various sectors of the digital economy. Their development and growing popularity underscore the crypto community's commitment to solving technical challenges while maintaining the decentralized ethos that lies at the heart of blockchain technology.

This journey through the most popular and influential cryptocurrencies, from Bitcoin to innovative Layer 2 solutions, highlights the sheer diversity and potential within this space. Each cryptocurrency, with its unique features and purposes, contributes to a broader, more inclusive financial ecosystem. However, as enticing as the opportunities in this field may seem, they come with their own set of complexities and risks.

It is crucial, therefore, to emphasize the importance of conducting thorough research before investing in any cryptocurrency. The volatility and regulatory uncertainties surrounding digital assets require prospective investors to be well-informed and cautious. Investing in cryptocurrencies is not merely about following trends or market sentiments; it involves understanding the underlying technologies, market dynamics, and potential legal implications.

In conclusion, the world of cryptocurrencies offers a fascinating glimpse into what the future of finance could look like. As you continue to explore this dynamic field, remember to approach it with curiosity, caution, and a commitment to learning. The decisions you make should be grounded in knowledge and an understanding of your own financial goals and risk tolerance. The journey into cryptocurrencies is not just an investment of your funds but an investment in your understanding of a rapidly evolving digital economy. Keep looking, keep asking questions, and let knowledge and wise decision-making guide you on your journey.

Section V
Looking Ahead

Chapter 14
Introducing Cryptonomics

As we delve into the fabric of modern economic theory, a novel paradigm emerges that challenges the foundational tenets of Mercantilism, Capitalism, and Keynesianism. This new framework, known as Cryptonomics, is more than a mere adjunct to existing economic theories—it represents a seismic shift in our understanding of value, exchange, and the mechanics of financial systems. Cryptonomics, at its essence, is the embodiment of a decentralized value system, a re-envisioning of economic principles through the lens of cryptography and blockchain technology.

The term Cryptonomics hijacks the traditional lexicon of economics, infusing it with the disruptive spirit of the digital age. It is a term that speaks to an economy where trust is built not on the reputation of institutions or the authority of governments but on the immutable laws of mathematics and the unassailable logic of algorithms. In the world of Cryptonomics, markets are not just free but also fair, not simply efficient but inherently transparent.

Cryptonomics transcends the concept of digital currencies. It encapsulates an economic philosophy where decentralization is the cornerstone. Unlike Mercantilism, which hoards wealth for national power, Cryptonomics disperses it, empowering individuals and communities. It stands apart from Capitalism's invisible hand, instead of placing the levers of economic power into the visible and verifiable hands of the collective. And where Keynesianism advocates for government intervention to stabilize markets,

Cryptonomics relies on the inherent stability of distributed consensus mechanisms.

In this new economic order, Cryptonomics challenges the role of fiat currencies and central banks. It proposes a world where money supply is not dictated by policy but algorithmically defined, where inflation is not a tool of statecraft but a function of network design. The theory of Cryptonomics sees the global economy not as a collection of national systems but as a unified, borderless marketplace, where value transfer is as simple and swift as the exchange of information.

The implications of adopting Cryptonomics are profound. It paves the way for a reimagined financial infrastructure—one where financial inclusion is intrinsic and economic agency is returned to the individual. With its roots in cryptographic security, it ensures that transactions are secure and privacy is protected, all while offering a new form of economic freedom and opportunity.

At the forefront of Cryptonomics lies the decentralization of power in economic systems. This principle challenges the conventional centralization of monetary authority, dispersing it across a network where individuals collectively make decisions and validate transactions. This diffusion of power reconfigures the economic landscape, creating a more democratic and equitable environment for financial engagement.

Trust in Cryptonomics is not derived from institutional reputation or governmental guarantee but is established through cryptographic verification. Each transaction within a blockchain is secured by algorithms that offer an unprecedented level of certainty. This cryptographic trust is fundamental, ensuring that economic exchanges are tamper-proof and verifiable by anyone, thus reducing the possibility of fraud and corruption.

The "Invisible Ledger," an evocative term for the blockchain, plays a critical role in market dynamics within Cryptonomics. It acts as an omnipresent, incorruptible arbiter of transactions and ownership. Unlike the "invisible hand" that Adam Smith envisioned guiding free markets through self-interest, the "Invisible Ledger" provides a transparent and permanent record of economic activity, accessible to all, ensuring fairness and efficiency.

Individual sovereignty over money is another pillar of Cryptonomics. It empowers users with absolute control over their digital assets, free from the risk of censorship or seizure by centralized entities. This sovereignty heralds a shift from the traditional model of trust in third parties to a system where individuals can exercise full autonomy over their economic choices.

The notion of value and scarcity in the digital realm is redefined through Cryptonomics. Unlike fiat currencies, which can be printed at will, digital currencies like Bitcoin have a predefined scarcity, echoing the properties of precious metals. This digital scarcity is carefully encoded within the fabric of the currency, imbuing it with intrinsic value and resilience against inflation.

Finally, the efficiency of markets within Cryptonomics is greatly enhanced. The frictionless nature of blockchain-based transactions, the elimination of intermediaries, and the automation of complex contracts lead to markets that are not only efficient but also inherently transparent and globally accessible.

Tokenization and the Expansion of Value

As we venture deeper into the realm of Cryptonomics, we encounter the revolutionary concept of tokenization, which extends the utility of cryptocurrencies far beyond the mere exchange of digital cash. Tokenization is the process of converting rights to an asset into a digital token on a blockchain. This paradigm shift has profound implications, as virtually anything of value can be tokenized — from tangible assets like real estate and art to intangible assets like intellectual property and voting rights.

The tokenization of assets marks a significant departure from traditional economic practices. It offers a new model of ownership and investment, one that is fractional, borderless, and easily transferable. Tokens represent a convergence of diverse asset classes into a unified digital form, enabling streamlined transactions over blockchain networks and opening up a global marketplace previously constrained by geographical and regulatory barriers.

The applications of tokens within the economy are diverse and continually expanding. Utility tokens can grant access to services or resources within a network, while security tokens can represent equity, shares, or interest in an organization or project. Tokens are also being used to facilitate crowd-sourced funding and investment through initial coin offerings (ICOs), allowing for the democratization of capital formation.

One of the most significant implications of tokenization is its potential to transform the way we interact with real-world assets. By tokenizing real estate, for instance, property ownership can be divided into tokens that represent fractions of the underlying asset, enabling individuals to own portions of property and benefit from rental income and appreciation. This not only makes investment more accessible but also enhances liquidity in markets that are traditionally illiquid.

Tokenizing real-world assets also introduces new efficiencies in asset management and transfer. Smart contracts can automate processes such as dividend distribution, voting rights, and buy-back options, reducing administrative overhead and the potential for human error. Furthermore, the transparency and immutability of blockchain ensure that the provenance and ownership of tokenized assets are always verifiable, mitigating the risk of fraud.

As tokenization continues to gain momentum, it challenges the very architecture of our current economic system. It proposes a world where the barriers between asset classes dissolve, where liquidity is a given, and where markets operate with unprecedented efficiency and inclusivity. The transition to a tokenized economy is not without its hurdles, including regulatory adaptation and market acceptance. Yet, the trajectory of tokenization suggests a compelling vision of the future — one where the expansion of value through digital tokens redefines our relationship with assets and opens new horizons for investment, ownership, and economic participation.

Cryptonomics and Monetary Policy

In the age of cryptocurrencies, traditional monetary policy, once the exclusive domain of central banks, faces a profound reimagining. Cryptonomics ushers in an era where the levers of economic control can be algorithmically programmed and executed without the need for central bank intervention. This shift from a centralized to a decentralized approach in monetary policy has profound implications for how economies could be managed in the future.

Algorithmic regulation represents the heart of this shift, offering a stark contrast to the discretionary policies often employed by central banks. Within the domain of Cryptonomics, the supply of money is not adjusted by committee decisions or economic projections but is instead governed by pre-set rules embedded within the currency's code. This automated approach to monetary policy provides a predictable, transparent, and tamper-proof method of regulation that stands in contrast to the sometimes opaque and reactive nature of traditional monetary policy.

The cryptonomic approach to inflation and deflation diverges significantly from conventional methods. Rather than actively managing the money supply to target inflation rates, cryptocurrencies like Bitcoin have a predetermined issuance schedule that mimics the scarcity and finite supply of precious metals. This algorithmic scarcity can potentially mitigate inflationary pressures endemic to fiat currencies, which can be printed at the will of governments, often leading to the devaluation of money.

Moreover, deflation, typically a concern for traditional economies due to its association with reduced consumer spending, may play a different role in a cryptonomic context. In an ecosystem where the value of currency can increase over time, the propensity to spend is balanced by the propensity to save, reflecting a shift in consumer behavior and challenging traditional economic theories around deflation.

Cryptonomics also presents a novel approach to managing the money supply. Instead of central banks that adjust monetary policy based on current economic conditions, the supply of cryptocurrencies is algorithmically fixed, introducing a level of predictability into the economic equation. This fixed supply schedule stands as a bulwark against the arbitrary expansion of the money supply and represents a new philosophy in the management of national and global economies.

As we consider the future of Cryptonomics and monetary policy, it is clear that the integration of blockchain technologies into the financial sector will necessitate a re-evaluation of long-held economic principles. With the cryptonomic model, we step into uncharted territory, exploring the potential for a more stable and predictable economic foundation built not on the shifting sands of policy change but on the solid ground of cryptographic certainty.

Capital Formation and Investment
The advent of Cryptonomics has propelled the mechanisms of capital formation into a new era with Initial Coin Offerings (ICOs) and Security Token Offerings (STOs), presenting a novel paradigm that challenges traditional securities laws. These digital fundraising methods, while

democratizing investment, inherently conflict with the established regulatory frameworks designed to protect investors and maintain fair, orderly, and efficient markets.

In the United States, the Securities and Exchange Commission (SEC) has been at the forefront of this conflict, asserting that many tokens issued through ICOs and STOs meet the criteria of securities and thus fall under their jurisdiction. The clash arises from the borderless nature of cryptocurrencies, which doesn't align neatly with national securities laws that are bound by territorial jurisdiction. The SEC has maintained that any offering deemed to be a security must comply with its stringent registration requirements unless a valid exemption applies, regardless of whether these tokens are transacted using blockchain technology.

This stance has led to a contentious atmosphere where some ICOs and STOs have faced legal action for failing to adhere to securities laws, resulting in penalties and, in some cases, the return of funds to investors. Such enforcement actions underline the SEC's commitment to protecting investors from the risks associated with unregulated offerings, which can include fraud and market manipulation.

Despite the friction, these developments have also fostered a dialogue between regulators and the cryptocurrency community. There is a growing recognition of the need for a regulatory framework that can accommodate the innovation of Cryptonomics while upholding the investor protections that securities laws are intended to provide. Some industry participants have proactively sought guidance from the SEC through formal inquiries and have attempted to structure their offerings to comply with existing securities regulations.

The resolution of this conflict is critical for the future of capital formation within Cryptonomics. A clear and consistent regulatory framework would not only provide legal certainty to issuers of digital tokens but also could enhance investor confidence in these novel investment vehicles. It's a delicate balance to strike—preserving the innovative spirit that fuels Cryptonomics while ensuring the stability and integrity of financial markets.

As case law evolves and regulatory stances are clarified, the hope is that a harmonious integration of Cryptonomics within national securities frameworks can be achieved. This would pave the way for a new chapter in the annals of capital formation, where the revolutionary potential of blockchain-based fundraising can be realized within the bounds of the law, ensuring a fair and secure investment landscape for all parties involved.

Rethinking Economic Cycles

Cryptonomics invites a re-evaluation of economic cycles as understood in traditional financial systems. In the realm of cryptocurrencies and blockchain, the forces that drive boom-and-bust cycles differ fundamentally from those of the conventional economy. Rather than being primarily influenced by monetary policy, interest rates, and fiscal stimuli, economic cycles within Cryptonomics are propelled by a unique set of drivers.

One of the principal drivers of economic cycles in Cryptonomics is the network effect. The value of a cryptocurrency or a blockchain platform often hinges on the number of users and the extent of its adoption. As more people join and utilize a network, its value and utility increase, potentially leading to rapid growth. This growth, however, can sometimes outpace the underlying fundamentals, leading to speculative bubbles and subsequent corrections, mirroring traditional economic cycles but driven by different underlying dynamics.

Technological adoption plays a central role in shaping the contours of economic cycles in Cryptonomics. Innovations such as smart contracts, decentralized finance (DeFi), and non-fungible tokens (NFTs) create new economic opportunities and markets virtually overnight. The pace of technological change and the market's response to these innovations can lead to periods of intense activity followed by stabilization as the technology matures and integrates into the broader economy.

Moreover, Cryptonomics holds the potential to reduce systemic risk within the economy. The decentralization inherent in blockchain technology means that risks are distributed across the network rather than concentrated in a few central institutions, as is often the case in traditional finance. This dispersion of risk can mitigate the domino effect seen in centralized systems where the failure of a single entity can lead to widespread economic fallout.

However, this potential for reduced systemic risk does not eliminate the possibility of volatility and cycles of expansion and contraction. The cryptonomic economy is still susceptible to the human behaviors of over-enthusiasm and irrational pessimism that characterize all markets. The difference lies in the underlying architecture of the system, which, by design, is more open and resilient.

As we continue to navigate the developing world of Cryptonomics, understanding the unique drivers of its economic cycles becomes paramount. With this knowledge, participants can better anticipate market movements and protect against the kind of systemic shocks that have historically plagued financial systems. In this way, Cryptonomics does not just offer a new set of economic tools, but it also provides a blueprint for a more stable and sustainable financial future.

Labor Markets and Employment

Cryptonomics heralds a transformation of labor markets and employment paradigms, redefining how work is performed, compensated, and valued. In a cryptonomic landscape, traditional employment models are giving way to more flexible, decentralized, and direct forms of economic participation.

The structure of work itself is being reshaped by the influence of Cryptonomics. The gig economy, already on the rise due to digital platforms, finds a natural ally in cryptocurrencies, which facilitate swift and secure transactions across borders. This synergy enables a seamless exchange of labor and compensation on a global scale. Freelancers can be paid instantly in digital currency for their services, regardless of their location, bypassing the delays and fees associated with traditional banking systems. This opens up international employment opportunities, allowing individuals to compete in a global marketplace and companies to tap into a vast pool of talent.

The compensation models within Cryptonomics also undergo significant transformation. With the advent of smart contracts, payments for work can be automated and linked to the completion of specific tasks or milestones, ensuring transparency and trust between parties. Furthermore, individuals can earn cryptocurrency through various means, such as participating in network validation (mining), contributing to open-source projects, or providing services within decentralized applications.

Empowerment is a central tenet in the cryptonomic vision for labor markets. Cryptocurrencies and blockchain technology provide individuals with the tools for direct economic participation, cutting out intermediaries that traditionally stand between a worker and their earnings. This direct participation extends to the ability to raise capital through mechanisms like Initial Coin Offerings (ICOs), enabling entrepreneurs and small businesses to fund their ventures and ideas through community support rather than institutional investment.

In Cryptonomics, the notion of work extends beyond the mere exchange of time for money. It encompasses a broader spectrum of value creation and exchange, where labor, creativity, and innovation are directly connected to economic outcomes. As these new models of work and compensation continue to evolve, they promise a future where employment is not only more accessible and equitable but also more closely aligned with the individual's contributions and capabilities.

The Global Economy and Cryptonomics

Cryptonomics is poised to redefine the fabric of the global economy through its potential to enhance global trade with frictionless transactions. The advent of blockchain technology and cryptocurrencies eliminates many of the traditional obstacles to economic exchange, such as lengthy settlement times, cumbersome paperwork, and the need for trusted intermediaries. Transactions that once took days to clear can now be settled in minutes, if not seconds, regardless of the distance or the amount, enabling a level of agility and efficiency previously unattainable in international trade.

Currency barriers and exchange rate inefficiencies have long been a thorn in the side of global commerce, introducing risk and unpredictability into cross-border transactions. Cryptonomics offers a salient solution to these issues by providing a universal medium of exchange that transcends national currencies. With cryptocurrencies, businesses can bypass the complexities of currency conversion, hedging against volatility and reducing costs. This streamlining of transactions paves the way for a more interconnected and fluid global marketplace, where the economic potential of regions can be unlocked without being hindered by the fiat currency ecosystem.

Perhaps the most profound impact of Cryptonomics on the global economy is its inclusivity and accessibility. By leveling the playing field, it allows individuals and businesses from developing regions to participate in the global economy on a more equal footing. Small enterprises can access international markets without the prohibitive costs of traditional banking services, and individuals can receive remittances or engage in trade without the need for a bank account. This inclusivity extends beyond mere access to financial systems; it empowers entrepreneurs and innovators in emerging economies to compete and collaborate on a global stage.

Cryptonomics does not merely streamline existing processes; it reimagines the potential of the global economy. It envisions a world where the barriers to economic exchange are not geographical borders or financial gatekeepers, but rather the limits of innovation and ambition. As Cryptonomics continues to gain traction, it promises to foster a more inclusive, efficient, and interconnected global economic landscape.

The Future of Cryptonomics

The future of Cryptonomics is not just a projection but a vibrant canvas of possibilities, as cryptoeconomic models and theories continue to evolve and intersect with the broader economic landscape. This evolution is marked by the exploration of new frameworks for value exchange, the development of sophisticated financial instruments on the blockchain, and the rigorous debate over the implications of a decentralized financial future.

Cryptonomics, with its radical rethinking of value and trust, is poised to integrate more deeply with traditional financial systems. This integration is a complex dance between innovation and regulation, between the disruptive potential of blockchain technology and the structured stability of established financial institutions. As these systems converge, we can expect to see a hybridization of services — where the speed, transparency, and inclusivity of Cryptonomics merge with the scale, security, and familiarity of traditional finance.

Forecasting the long-term impacts of Cryptonomics on society and the global economy requires both imagination and caution. The promise of financial democratization and global economic participation is profound, with the potential to reduce income inequality, enhance accessibility to capital, and create new economic opportunities. Yet, the broader societal implications — from shifts in employment models to changes in government fiscal policies — are vast and as yet not fully understood.

As we conclude, it is evident that Cryptonomics is more than a technological phenomenon; it is an economic revolution that is reshaping the very concept of finance. Its principles challenge us to reimagine the exchange of value in a world where trust is established through consensus and cryptography, not central authority. The role of Cryptonomics in shaping the future of finance is indelible — it stands as both a beacon of innovation and a testament to the enduring quest for a more equitable and efficient economic system. In this unfolding narrative, the fusion of technology and economics is crafting a new epoch of financial empowerment, promising a future where the true potential of global economic collaboration can be realized.

Chapter 15
Bitcoin and Blockchain: Remedies for an Ailing Economic System

As we embark on this chapter, it is essential to cast a reflective eye on the journey thus far. The preceding chapters laid bare the economic maladies that have silently festered within the traditional systems of trade and value. From the mercantilist hoarding of wealth that stymied global economic participation to the Keynesian reliance on government intervention that sometimes led to further economic distortions, we dissected a series of inherent flaws. These models, once revolutionary, now grapple with challenges in a world that has rapidly outgrown their confines.

Yet, amidst these critiques, a new epoch is emerging. In the wake of these systemic vulnerabilities, Bitcoin and blockchain technology step forward, heralding not just incremental adjustments but a fundamental rethinking of economic principles. Bitcoin, in its essence, is the antithesis of centralized control, offering an immutable ledger that stands as a testament to every transaction. Blockchain, the bedrock upon which Bitcoin rests, is more than a technology; it is a new paradigm of trust, where consensus drives validation and every participant holds a stake in the integrity of the whole.

As we delve deeper into this chapter, we will explore how the decentralized doctrine of Bitcoin and the innovative prowess of blockchain provide promising avenues for rectifying the economic ailments detailed earlier. This is not a mere academic exercise but a timely and critical discourse as we stand at the cusp of what could be the most significant economic revolution since the inception of fiat currency. The coming pages are an invitation to witness how Bitcoin and blockchain could redefine the very scaffolding of financial interaction, heralding an era of empowerment, stability, and inclusivity previously unimagined.

The Unbanked

In the global economy, a staggering number of individuals remain outside the traditional banking system. Estimates suggest that nearly 1.7 billion adults worldwide are unbanked, lacking access to a bank account or the means to secure their financial future through savings, credit, or insurance. This figure represents a substantial portion of the global population, constrained not by choice but by circumstance, from participating in the formal financial economy.

Bitcoin emerges as a beacon of hope in this context. It offers a solution that is as elegant as it is revolutionary: a decentralized financial system where anyone with a mobile device can participate. With Bitcoin, the need for traditional banking infrastructure, with its attendant costs and barriers, dissipates. Instead, the unbanked can leapfrog directly to a financial system that is open, borderless, and inclusive.

The process is simple and straightforward. To engage with Bitcoin, all one needs is a smartphone and an internet connection. From there, setting up a Bitcoin wallet is free and requires none of the paperwork, minimum balance requirements, or service fees that come with traditional bank accounts. This ease of access is revolutionary, allowing individuals to send and receive money across borders with minimal fees, invest in savings that are not subject to the inflationary whims of local currencies, and even access credit through peer-to-peer lending platforms.

Bitcoin's potential to solve the problem of financial exclusion is not just theoretical. In countries battered by currency devaluation and hyperinflation, Bitcoin has already become a lifeline. It provides a stable store of value and a means of transaction when local currencies falter. Moreover, Bitcoin's underlying technology, the blockchain, ensures that all transactions are secure, transparent, and immutable. This security is paramount for the unbanked, who often operate in environments where financial fraud and instability are prevalent risks.

The Economic Flaws of Centralized Control
The modern financial system is a web of centralized institutions, from commercial banks to central banks, that control monetary policy and the flow of capital. This centralization comes with several pitfalls: bureaucracies slow to adapt to change, systems vulnerable to single points of failure, and the concentration of power that can lead to systemic corruption and inefficiency. Moreover, centralized control often means that individuals have a limited say over their financial choices, subject to the whims of institutions and the gatekeepers of finance.

Bitcoin challenges this paradigm by offering a decentralized alternative. At its core, Bitcoin is a peer-to-peer network that operates without a central authority. Transactions are verified by network participants—miners—who use computational power to solve cryptographic puzzles. Once a transaction is verified, it is added to the blockchain, a distributed ledger that is immutable and transparent. This decentralization ensures that no single entity has control over the Bitcoin network, making it resistant to censorship, immune to arbitrary monetary policy, and less susceptible to the risks that come with centralized control.

The benefits of Bitcoin's decentralized nature are manifold. Without the need for intermediaries, transactions can be faster and more cost-effective. The blockchain's transparency means that all transactions are publicly verifiable, fostering trust through openness rather than authority. The security provided by decentralization also means that Bitcoin is resilient against attacks that would typically cripple centralized systems, such as data breaches or bank runs.

Several case studies underscore the advantages of decentralization. For instance, in countries with hyperinflation or monetary instability, Bitcoin has provided a more stable alternative to local currencies. In Venezuela, where the Bolivar has suffered from rampant inflation, Bitcoin has become a means to preserve savings and conduct business. In Zimbabwe, after the collapse of the national currency, Bitcoin traded at a premium, reflecting its value as a haven in times of economic crisis.

Moreover, Bitcoin's decentralized nature has spurred innovations in financial services, such as decentralized finance (DeFi) platforms that operate autonomously, providing lending, borrowing, and trading services without the need for traditional financial institutions. These platforms have shown remarkable efficiency and growth, with billions of dollars locked in smart contracts, demonstrating the demand for decentralized financial services. In essence, Bitcoin's decentralization represents not just a technical feature but a philosophical shift towards empowering individuals.

Inflation and Fiat Currency Debasement

Inflation and currency debasement have been perennial problems of fiat currencies, where the value of money erodes over time, primarily due to an increasing supply. Central banks can print more money in response to various economic pressures, leading to inflation, which diminishes the purchasing power of the currency. This situation has historically led to the devaluation of savings and uncertainty in financial planning for both individuals and businesses.

Bitcoin presents a compelling countermeasure to this issue with its deflationary model. Unlike fiat currencies, Bitcoin has a fixed supply cap of 21 million coins, which is hardcoded into its protocol. This limited supply mimics the scarcity of precious resources like gold and creates a natural hedge against inflation. As no more bitcoins can be created once the cap is reached, the currency is protected from the kind of dilution that plagues fiat currencies, ensuring that it cannot be debased by any central authority.

The rate at which new bitcoins are introduced to the market is determined by the mining process. Approximately every four years, an event known as the halving cuts the rewards for mining new blocks on the blockchain by half. This halving process slows the rate at which new bitcoins are created and is a key component of Bitcoin's deflationary stance. It is designed to gradually reduce the supply of new bitcoins entering the market until the last bitcoin is mined around the year 2140. This contrasts starkly with the expansionary monetary policies often adopted by central banks around the world.

Moreover, the transparent and predictable issuance of Bitcoin provides a level of certainty in an economic landscape often clouded by complex and opaque monetary policies. For those living in countries with hyperinflation or unstable monetary regimes, Bitcoin offers a potential lifeline and a store of value that stands apart from their domestic currencies.

Financial Crises and Systemic Failures
Financial crises and systemic failures have often exposed the frailties of our traditional financial system—fraught with over-leveraged institutions, lack of transparency, and the domino effect of interconnectedness. When one pillar falls, it threatens to bring down the entire structure. The global financial crisis of 2008 stands as a stark reminder of these vulnerabilities.

Bitcoin and the blockchain technology that underpins it, offers innovative solutions to these systemic issues. Its decentralized nature means that the system does not hinge on the solvency of banks or the stability of governments. Instead, it is supported by a distributed network of nodes, each verifying transactions independently. This reduces the systemic risk associated with central points of failure, making the financial system more resilient to individual shocks.

In times of crisis, Bitcoin could serve as a haven for capital, akin to gold. Its decoupled nature from the traditional financial ecosystem allows it to operate independently of market pressures that typically lead to liquidity crunches. During the economic turmoil caused by the COVID-19 pandemic, for example, Bitcoin's price increased significantly, suggesting its growing perceived role as a "digital gold" that can store value in tumultuous times.

Moreover, blockchain's inherent transparency provides a real-time, immutable ledger of all transactions, fostering greater trust in the system. This transparency can prevent the type of risky financial practices that lead to crises, as all transactions are open to verification by any participant in the network. The opacity that allowed for the accumulation of toxic assets leading up to the 2008 crisis would be far less feasible in a blockchain-based system.

The diversification of risk is another critical feature of Bitcoin and blockchain. By enabling the tokenization of assets, blockchain allows for the fractional ownership of assets, from real estate to art, making it easier to spread risk across a wider array of investments. This democratization of investment opportunities can lead to a more stable financial ecosystem, as it lessens the impact any one asset or asset class can have on an individual's portfolio.

Furthermore, cryptocurrencies can strengthen financial systems by providing new forms of liquidity and mechanisms for raising capital. Initial Coin Offerings (ICOs) and Security Token Offerings (STOs) have already shown how blockchain can be used to fund ventures directly from investors around the world, bypassing traditional capital-raising methods and their associated systemic risks.

In essence, Bitcoin and blockchain present a paradigm shift from a centralized to a decentralized financial framework, one in which the power is distributed among its users rather than concentrated in a few entities. While not a panacea, the adoption of cryptocurrencies holds the potential to mitigate the impact of financial crises and systemic failures, providing a foundation for a more robust, transparent, and diversified global financial system.

Censorship Resistance

Traditional financial systems, with their intricate web of regulations and reliance on central authorities, can sometimes lead to situations where individuals and entities face financial censorship and the risk of asset seizure. Governments can freeze bank accounts, international sanctions can prevent cross-border transactions, and political unrest can lead to the confiscation of assets, leaving individuals powerless and without recourse.

Bitcoin, built on a decentralized blockchain, provides a robust solution to these challenges. Its decentralized nature means there is no central point of control that can be leveraged to censor transactions or seize assets. Ownership of Bitcoin is established through cryptographic keys; only the person with the private key can access and transfer their Bitcoin, granting users a level of financial autonomy that is nearly impossible to achieve with traditional currencies.

This sovereignty is further enforced by Bitcoin's global nature. It transcends national borders and regulatory jurisdictions, enabling individuals to engage in financial transactions with peers anywhere in the world. Bitcoin's censorship-resistant qualities have made it particularly valuable for individuals in oppressive regimes or unstable economies. For instance, in countries where freedom of expression is curtailed, activists have turned to Bitcoin to receive funding securely. Similarly, in nations experiencing hyperinflation or aggressive financial controls, Bitcoin has become a means to preserve wealth and make purchases without government interference.

Real-world examples of Bitcoin safeguarding individual wealth are numerous and telling. During the Cypriot financial crisis in 2013, when the government imposed strict capital controls and took a portion of deposits to recapitalize banks, many turned to Bitcoin to protect their assets from seizure. In Venezuela, faced with hyperinflation and a collapsing economy, citizens have increasingly adopted Bitcoin as a way to store value and conduct transactions, as the Bolivar's value plummeted.

Moreover, Bitcoin has been instrumental in humanitarian efforts where traditional banking systems fail. In conflict zones or after natural disasters, when banking infrastructure is compromised, Bitcoin has enabled the transfer of funds to assist those in need quickly and without the need for physical transport, which can be dangerous or impractical.

In essence, Bitcoin's design inherently includes mechanisms for ensuring financial sovereignty and resistance to censorship, providing a powerful tool for individuals to maintain control over their financial resources. As the digital currency continues to mature, its role in protecting and empowering users in the face of censorship and asset seizure will likely become even more pronounced, showcasing the potential for a financial system that upholds the values of autonomy and freedom.

Can Crypto Fix Corporatism?

Corporatism—a system where corporate groups are given priority over individuals, often leading to a concentration of power and wealth—is a complex societal and economic issue. It's characterized by the perpetuity of corporations (eternality), where their existence extends beyond the lives of their founders and can theoretically continue indefinitely. This often results in a relentless pursuit of profit (greed motive), sometimes at the expense of broader societal and environmental interests (the commons). Moreover, the close relationship between corporations and government (undue influence) can lead to regulatory capture, where businesses manipulate policies in their favor, further entrenching their power.

Bitcoin and blockchain technology offer potential solutions to some aspects of corporatism by enabling more decentralized and transparent systems:

Decentralization Against Eternality: Corporations, as immortal entities, can accumulate vast resources and influence over time. Blockchain introduces the possibility of more decentralized business models, such as Decentralized Autonomous Organizations (DAOs), which operate on blockchain principles and distribute decision-making among their members. This could mitigate the risks associated with the concentration of power in perpetuity.

Aligning Incentives with the Commons: The greed motive inherent in shareholder-driven corporations often sidelines the commons. Bitcoin's underlying technology can facilitate the creation of more complex incentive structures through smart contracts that can programmatically include considerations for social and environmental impact, tying business success to positive contributions to the commons.

Reducing Undue Governmental Influence:
Blockchain's transparent and immutable ledger can help diminish undue influence by making financial transactions and contributions transparent. This can potentially reduce corruption and increase accountability in political funding and governmental contracts.

While Bitcoin and blockchain have the potential to address certain aspects of corporatism, its important to recognize that technology alone cannot completely fix systemic socio-economic issues. The transition to blockchain-based systems would require significant changes in legal frameworks, corporate governance structures, and societal norms. Additionally, the risk of new forms of centralized power emerging within the blockchain ecosystem itself should not be overlooked.

Governments Hoarding Wealth (Mercantilism)

Bitcoin's emergence as a digital asset introduces a nuanced counter to traditional mercantilist practices, notably in how governments accumulate wealth and manage their currencies for trade surpluses. Firstly, while governments could theoretically attempt to stockpile Bitcoin as a digital reserve, the cryptocurrency's inherent design imposes limits on such behavior. No single entity can control or manipulate the Bitcoin network, which inherently curtails the impact any government's hoarding could have on the broader economic system.

Secondly, the fixed supply of Bitcoin—capped at 21 million—naturally resists the inflationary tactics typically employed in mercantilist strategies. Governments cannot produce more Bitcoin to inflate their way out of debt or to devalue their currency to gain export advantages. This fixed supply, akin to the gold standard, could serve as a check against the devaluation of the currency in the global trade arena, fostering a more stable economic environment that is less susceptible to manipulation by any single nation.

In conclusion, Cryptonomics holds the promise of potentially revolutionary solutions to the shortcomings of previous economic theories. It offers a fresh paradigm, one that is built on the principles of decentralization, transparency, and immutability. As a discipline at the intersection of cryptography and economics, it extends beyond the boundaries of traditional financial systems, challenging long-established practices and offering new ways to consider value, wealth, and monetary policy.

Where previous economic models have often relied on centralization and control, Cryptonomics empowers individuals with greater control over their financial destinies. It mitigates the risks of inflation and currency debasement that plague fiat systems and introduces a model of currency that is not subject to the whims of governments and financial institutions.

The rise of Cryptonomics, spearheaded by the advent of Bitcoin and other digital assets, could signal a significant shift in how we understand and interact with money. While it is not without its challenges and areas for development, the foundational concepts of Cryptonomics present compelling alternatives to the economic systems of the past, potentially heralding a more equitable and efficient global economy. As we navigate its implications and integrate its innovations, Cryptonomics may well redefine the economic landscape of the future.

Chapter 16
The Potential of a Decentralized Future.

As we delve deeper into the potential of a decentralized future, we encounter one of the most intriguing developments of the blockchain era: Decentralized Autonomous Organizations, or DAOs. These entities represent the crystallization of decentralized governance, where the traditional hierarchical management structures are replaced by a set of self-executing rules on a blockchain. DAOs are entities without a central leadership, managed by members who make decisions collectively, and not through a central figure or a management team. The very essence of DAOs is encoded in smart contracts that lay out the rules of engagement and automatically enforce them.

The history of DAOs is a narrative of ambition, learning, and evolution. From the pioneering days of the first DAOs, which were ambitious but fraught with security challenges, as exemplified by the infamous DAO incident in 2016 where a significant amount of cryptocurrency was siphoned due to code vulnerabilities, the ecosystem has learned and adapted. This event served as a watershed moment, signaling the need for improved security measures and more rigorous development standards. Despite such setbacks, the resilience of the concept is evident in the myriad of successful DAOs that have since emerged, each learning from the lessons of the past.

Today, DAOs are not just theoretical constructs but functioning entities that are increasingly being integrated into various sectors. They have found applications in investment, where group funds are managed collectively; in content creation, where creators and consumers share ownership and profits; and even in charity, where transparency and accountability are paramount. Case studies of successful DAOs highlight their potential for fostering collaboration, streamlining decision-making, and democratizing access to investment opportunities. Across these diverse sectors, DAOs are increasingly embraced as they exemplify the principles of transparency, inclusivity, and collective decision-making, heralding a new chapter in organizational structures and potentially redefining the future of work, governance, and collective enterprise.

In the evolving tapestry of organizational structures, DAOs emerge as a potential replacement for traditional corporations, particularly in their ability to address the pitfalls of corporate centralization. Traditional corporate models, while having stood the test of time, often suffer from issues like bureaucratic inertia, lack of transparency, and the concentration of decision-making power. Such centralization can stifle innovation and responsiveness, and at times, alienate stakeholders from the decision-making process.

DAOs offer a compelling contrast by embodying agility and efficiency. With decisions made through consensus mechanisms rather than top-down directives, DAOs can adapt more rapidly to changing market conditions and technological advancements. This fluidity is not just about speed; it's about the organic alignment of an organization's direction with the collective will of its members. The agility of DAOs comes from their very nature—being software-driven and community-governed—allowing them to operate with minimal bureaucracy.

Shareholder democracy in DAOs is not merely a regulatory requirement but a foundational characteristic. Each member, or token holder, has a stake in the governance of the organization, with the power to propose or vote on decisions directly. This participatory model fosters a more democratic form of corporate governance, where the disparity between ownership and control found in traditional corporations can be significantly reduced.

In DAOs, the ethos of shareholder democracy is realized more fully because every stakeholder's voice can be heard and counted. There is a direct line from ownership to action, a democratized approach to corporate governance that can potentially reshape how companies are run and how they interact with their communities of users, investors, and advocates. Such a shift could herald a future where companies are truly accountable to those they serve, operating with heightened transparency and alignment with the collective interests of their stakeholders.

The advent of DAOs has not only revolutionized corporate governance but also posits a transformative potential for governmental systems. In the theoretical models where DAOs underpin governance, the traditional layers of bureaucracy could be supplanted by blockchain-enabled platforms that allow for direct participation and decision-making by citizens. These models propose a form of government where policies and regulations are not only made by the representatives but also by the direct input of the populace, whose votes and opinions are recorded transparently on a blockchain.

The benefits of such a transparent and participatory approach to governance are manifold. It promises to enhance the democratic process by eliminating intermediaries, reducing the risk of corruption, and ensuring that the will of the people is directly reflected in governmental actions. With every transaction and vote being recorded on an immutable ledger, citizens can see the direct impact of their participation, fostering a greater sense of involvement and trust in the political process.

However, the application of DAOs within governmental systems is not without significant challenges and considerations. The transition from traditional governance structures to DAO-based models requires careful deliberation. Questions of security, privacy, and the digital divide come to the forefront. There is also the matter of ensuring that the algorithms and code underpinning DAOs are fair, unbiased, and represent the diversity of the population they serve.

Moreover, the shift towards DAOs in governance would necessitate a cultural change, where citizens are not only allowed but also expected to engage directly with and take responsibility for governance. It requires robust education and familiarization with the technology, ensuring that all members of society can participate equally. Legal frameworks would also need to evolve to accommodate this new form of governance, addressing issues such as liability, dispute resolution, and the enforcement of DAO-made decisions.

DAOs present a vision of governance that is more fluid, transparent, and participatory than ever before. Yet, the journey toward realizing this vision will be iterative, requiring careful consideration of the complexities of governing diverse and dynamic societies. The promise is there: a government system that is more responsive to its citizens and better reflects their collective will, but the path to that future must be trodden with diligence and an eye toward the lessons learned from every step forward.

As we transition from the domain of governance to the realm of economics, we witness the ascent of Decentralized Finance, commonly known as DeFi - a system that seeks to recreate and redefine traditional financial services with no central authority in charge. At its core, DeFi is an amalgamation of blockchain technology, digital assets, and smart contracts that together forge an open financial ecosystem. This ecosystem is characterized by its inclusivity, offering unhindered access to financial services for anyone with an internet connection, irrespective of geography.

The building blocks of DeFi are diverse and innovative, ranging from protocols that enable lending and borrowing without intermediaries to platforms that allow for complex financial instruments like derivatives and yield farming. These services, which traditionally required a centralized authority or intermediary, are now executed on transparent, permissionless networks.

The current DeFi landscape is a testament to rapid growth and adoption, drawing users with the allure of greater control over their financial dealings and the promise of higher yields than those found in conventional banking. Success stories abound, from platforms that have locked in billions of dollars worth of assets to protocols that have enabled new forms of asset exchange and investment strategies. These milestones mark significant shifts in the perception and utilization of financial services.

Looking to the future, DeFi stands poised to further entrench itself in the global financial system. The opportunities for expansion and integration seem boundless, with the potential to bridge the gap between traditional finance and blockchain-based systems. Innovations in the space continue to emerge at a blistering pace, promising enhanced efficiency, lower costs, and greater access to financial services.

The projected trends suggest a blurring of lines between the old and new, where decentralized applications may run parallel to or even in collaboration with traditional financial institutions. We are on the cusp of a financial revolution—one where the foundational tenets of trust, transparency, and participation are not just idealistic concepts but realizable features of a new economic landscape shaped by the principles of Decentralized Finance.

Amidst the burgeoning field of decentralized finance, another innovative concept has captured the public imagination: Non-Fungible Tokens, or NFTs. These unique digital assets represent ownership of a specific item or piece of content, such as art, music, or even tweets, backed by blockchain technology. Unlike cryptocurrencies like Bitcoin, where each unit is identical and interchangeable, NFTs are distinct and cannot be exchanged on a one-to-one basis, hence the term 'non-fungible'. This uniqueness and the ability to verify authenticity and provenance through the blockchain lend NFTs their value and appeal.

The technology behind NFTs allows for the creation of digital scarcity, a concept previously elusive in the digital realm where reproduction and distribution were as simple as a copy-and-paste command. With NFTs, digital ownership mimics the exclusivity of owning a physical object—say, an original painting—bringing the same sense of possession to the digital world. This digital ownership is secured and recorded on the blockchain, making it permanent and auditable.

NFTs have already made a significant impact on the creative industries. Artists, musicians, and content creators have found in NFTs a new medium for monetizing their work. They allow creators to sell their art directly to a global audience without the need for traditional intermediaries, and with the ability to program royalties for future sales directly into the token. The art world has seen record-breaking sales, and entertainers have embraced the technology to offer unique experiences and collectibles to their fans.

The potential applications of NFTs extend well beyond art and entertainment. They are beginning to make inroads into other markets and asset classes, such as real estate and intellectual property. NFTs can represent fractional ownership of physical properties or secure the rights to digital creations, offering a new way to buy, sell, and manage assets.

Looking forward, the intersection of NFTs with asset and ownership tracking could redefine concepts of ownership and identity in the digital age. They could be used to securely record titles to real-world properties, patents, and other legal documents, ensuring authenticity and reducing fraud. Moreover, as we grapple with the concept of digital identity, NFTs could provide a way to establish and verify individual identities online, with far-reaching implications for privacy, security, and social interaction.

NFTs embody the promise of blockchain technology to create a world where digital ownership is as concrete and verifiable as owning something you can hold in your hands. As they evolve, NFTs could reshape entire industries, redefine ownership, and open up new possibilities for how we interact with the tangible and intangible assets of our world.

In the tapestry of blockchain's innovations, the convergence of DAOs, DeFi, and NFTs is creating a synergy that extends beyond their individual capabilities. Together, they are more than just the sum of their parts; they represent a cohesive ecosystem that is reshaping traditional economic and social structures.

DAOs bring decentralized governance, allowing for collective decision-making and the democratization of organizational and financial control. DeFi reimagines finance, stripping away intermediaries to enable peer-to-peer transactions and innovative financial instruments. NFTs introduce uniqueness and proof of ownership to the digital world, providing a bridge between physical and virtual assets.

The integration of these decentralized innovations has the potential to disrupt the way we engage with money, art, and each other. For instance, a DAO might govern a DeFi protocol that issues loans against NFTs as collateral, creating a closed-loop system that exemplifies the power of decentralized finance and governance. This could revolutionize not only art and collectibles markets but also real estate and intellectual property rights, allowing for fractional ownership and investment in assets that were previously illiquid or inaccessible.

The impact on traditional economic and social structures is profound. Decentralized systems can reduce reliance on central banks and financial institutions, mitigate censorship, and promote financial inclusion. NFTs are already challenging the norms in art and entertainment, empowering creators and altering the dynamics of ownership and profit-sharing.

The path toward a comprehensive decentralized ecosystem requires technological maturation, user education, and regulatory clarity. As these technologies develop, they must remain user-friendly and accessible to ensure widespread adoption. Moreover, the ecosystem must be resilient, capable of withstanding security threats and scaling to accommodate a growing user base.

This journey toward a fully-fledged decentralized ecosystem is not without challenges, but the potential rewards are vast. A world where financial systems are open, transparent, and equitable; where creators are fairly compensated and have direct relationships with their audiences; and where every individual has a say in the systems that govern their digital and economic lives is a compelling vision—one that DAOs, DeFi, and NFTs are making a tangible reality.

The vision of a decentralized future stands as a beacon of potential transformation. This vision, driven by the innovations of blockchain technology, is not merely a series of incremental changes but a fundamental reimagining of economic and social systems. The decentralized ledger's promise is nothing short of an economic revolution, where power is distributed, access is democratized, and transparency is the norm.

The journey toward this future is as exhilarating as it is uncertain. Traditional institutions are being challenged to adapt or reinvent themselves, and individuals are being empowered to take control of their financial destinies. The emerging decentralized landscape brings with it a host of unknowns: regulatory responses, technological advancements, and shifts in societal attitudes. Navigating this uncertainty requires a spirit of adaptability and a willingness to embrace change, knowing that the path forward may not always be linear or clear.

Yet, in this era of transformation, the most significant role is played by individuals and communities. It is the collective action of these stakeholders that drives adoption, shapes governance, and infuses these systems with value. Communities are forming around DAOs, DeFi protocols, and NFT marketplaces, each contributing to the robustness and vibrancy of the ecosystem. Individuals are not just passive participants; they are builders, creators, and decision-makers, actively shaping the future with each transaction, vote, or innovation.

As we stand on the cusp of what could be a seismic shift in the way we understand and interact with money, governance, and ownership, the decentralized future invites us to be both participants and architects. The potential of blockchain technology extends far beyond financial transactions; it offers a blueprint for a future where agency, inclusivity, and fairness are embedded in the very fabric of society. It is a future that we have the opportunity to build together, a decentralized future that beckons with the promise of a new economic dawn.

As the narrative of a decentralized future unfolds, it beckons each of us to become active participants in this burgeoning economy. This call to action is not just a summons to invest or engage in transactions; it is an invitation to contribute to a movement that is redefining the essence of economic interaction and governance. The decentralized economy thrives on participation—each new user, each additional transaction, and every smart contract adds to the strength and resilience of the network.

Moreover, advocacy for education in this new financial paradigm is critical. Knowledge is the currency of participation, and as the complexity of these systems grows, so too does the need for comprehensive education. This education must be inclusive, ensuring that individuals from all walks of life have the opportunity to learn about, engage with, and benefit from decentralized finance and the broader ecosystem of blockchain technologies.

But participation and education are only parts of the equation. The ongoing development and governance of decentralized systems demand active contributions from its community. Whether it's by developing new applications, participating in the governance of DAOs, or simply by providing constructive feedback to developers and policymakers, every contribution helps shape the future of these systems. It is through the collective input and collaboration of its users that the decentralized economy will continue to evolve, adapt, and innovate.

This call to action is not just for the technologically savvy or the financially literate—it's a call to all who envision a future where financial systems are transparent, accessible, and equitable. It's a call to those who see the potential for technology to empower and liberate rather than to control and confine. By stepping into this new realm, we each have the chance to leave our mark on the financial landscape of tomorrow, building a legacy of innovation, inclusion, and economic empowerment.

Chapter 17
A Multi-Crypto Future

In this chapter, we delve into the diverse and vibrant landscape of cryptocurrencies, extending our gaze beyond the pioneering Bitcoin to a broader ecosystem teeming with innovation and variety. This realm of digital currencies is not a monolith but a kaleidoscope, each crypto bringing its own unique features, purposes, and communities.

At the heart of this diversity is the understanding that the crypto world is much more than just Bitcoin. While Bitcoin may have laid the foundation, the ensuing years have seen the emergence of a multitude of cryptocurrencies, each designed with specific goals and functionalities. From Ethereum with its smart contract capabilities to Ripple's focus on fast and efficient cross-border payments, the range is extensive and continually expanding.

These cryptocurrencies differ not only in their technical design but also in their underlying philosophies and intended uses. Some, like Litecoin and Bitcoin Cash, were born out of the desire to improve upon Bitcoin's original model, offering faster transaction speeds or different scaling solutions. Others, like Cardano or Polkadot, aim to provide more robust and flexible platforms for decentralized applications and cross-chain interoperability.

Moreover, each cryptocurrency is often backed by a community of supporters and developers who believe in its particular vision. These communities are not just passive observers but active participants in the evolution of their chosen crypto, contributing to its development, governance, and adoption.

As we explore this varied landscape, it becomes evident that the world of cryptocurrencies is a testament to human ingenuity and the quest for financial innovation. It's a space where technology meets creativity, resulting in a spectrum of solutions each aiming to address different challenges and opportunities within the digital economy. This diversity is not just a strength but also a sign of the crypto world's maturity, indicating a future where multiple digital currencies coexist, each serving specific purposes within a broader, interconnected financial system.

As we venture deeper into the multi-crypto future, the potential for the primary function of Bitcoin emerges as a currency for intergovernmental trade and settlement. This idea represents a paradigm shift in how nations could engage in international trade, moving beyond the confines of traditional fiat currencies.

Bitcoin, with its decentralized nature, offers a compelling alternative for international trade. It operates independently of any single government or central bank, making it an attractive option for settling trade balances. This neutrality of Bitcoin could reduce the reliance on major reserve currencies like the U.S. dollar, potentially rebalancing global economic power dynamics.

One of the key advantages of using Bitcoin in governmental transactions is its ability to bypass the complexities and costs associated with traditional fiat currencies. Cross-border transactions with fiat can be slow and involve multiple intermediaries, each adding layers of fees and potential points of failure. Bitcoin transactions, on the other hand, can be completed faster and more efficiently, as they are processed on a peer-to-peer basis without the need for intermediaries.

Furthermore, Bitcoin's inherent characteristics, such as its fixed supply and resistance to censorship, make it a potentially stable and secure option for international trade. Unlike fiat currencies, which can be subject to inflation and political manipulation, Bitcoin offers a degree of predictability and security that could be highly appealing to governments looking to safeguard their international reserves.

In sum, the prospect of Bitcoin as an intergovernmental trade currency opens up new avenues for how nations interact economically. While there are challenges and uncertainties in its adoption at such a scale, the potential benefits it offers make it a serious consideration for the future in the evolving landscape of global finance.

In the tapestry of our multi-crypto future, the realm of everyday commerce stands as a vibrant and dynamic frontier, teeming with possibilities and challenges. As we peer into this future, the role of cryptocurrencies in daily transactions emerges as a pivotal theme, with altcoins like Dogecoin leading a fascinating narrative.

Altcoins for Daily Transactions

Imagine a world where, alongside traditional currencies, a plethora of digital coins jostle for space in our wallets – virtual wallets, that is. Dogecoin, once a whimsical creation, has found a place in this new world. Its journey from meme to mainstream highlights the evolving nature of cryptocurrencies. It's not just about Bitcoin anymore; altcoins, each with distinct characteristics, are vying for attention. Dogecoin, with its friendly mascot and a community-driven ethos, has carved out a niche for small-scale, routine transactions.

In this future, other altcoins too find their utility. Some, optimized for speed, make buying a coffee as quick as a swipe of a card. Others, boasting enhanced privacy features, become the go-to for discreet purchases. This diverse crypto landscape caters to varied transactional needs, reflecting the multifaceted nature of consumer behavior.

Analyzing Stability, Supply, and Inflation Rates

In envisioning the future of everyday commerce with altcoins, we must consider crucial economic aspects like stability, supply, and inflation rates, which are pivotal for their mainstream adoption. The supply dynamics of these cryptocurrencies, particularly Dogecoin, play a significant role in their economic viability for daily transactions. Distinct from Bitcoin's capped supply, Dogecoin adopts an inflationary approach, continuously introducing new coins into circulation. This model could encourage a culture of spending over hoarding, aligning well with the transactional nature of everyday purchases. However, it also brings into question the long-term sustainability in terms of value retention and price stability.

As we project into the future, the inflation rate of Dogecoin, while initially higher due to its ongoing coin production, is expected to taper down. Over time, as the total supply of Dogecoin increases, the percentage increased every year due to new coin minting becomes smaller relative to the existing supply. This gradual decrease leads us toward an inflation rate that could approximate 2%, a figure often cited as ideal in traditional economic theory for matching overall economic growth. This rate of inflation could strike a balance, offering a more stable value while still encouraging the use of Dogecoin for everyday transactions, making it a viable and potentially attractive option in the diverse landscape of cryptocurrency-based commerce.

Merchant Adoption and Consumer Preferences

The keystone in this crypto-commerce future is merchant adoption. The willingness of businesses, from the corner store to multinational conglomerates, to accept cryptocurrencies is crucial. This decision is influenced by factors like transaction speed, fee structures, and the volatility of these digital currencies.

As merchants weigh the benefits of adopting cryptocurrencies – lower transaction fees, access to a broader customer base, and reduced fraud risk – they also face challenges. The fluctuating value of cryptocurrencies can be a double-edged sword, promising gains but also posing risks.

Meanwhile, consumer preferences play a pivotal role. In this future, we see a generation more attuned to digital currencies, their spending habits shaped by the convenience and novelty of cryptos. Consumers revel in the choice and flexibility offered by these digital currencies, using Dogecoin for small, routine purchases, and perhaps turning to Bitcoin or stablecoins for more significant expenditures.

The Role Cryptocurrencies for Privacy

In our future cryptocurrency landscape, privacy coins like Monero stand out for their emphasis on anonymity and financial privacy. These coins are designed to provide users with a level of privacy that mainstream cryptocurrencies like Bitcoin and Ethereum do not.

Monero, a leading privacy coin, is crafted to obscure the details of the transaction parties and amounts involved. Unlike Bitcoin, where transactions are transparent and traceable on the blockchain, Monero employs sophisticated cryptographic techniques to ensure that transactions are not just secure, but also private. This privacy feature appeals to users who prioritize confidentiality, ranging from individuals seeking protection from surveillance and data breaches to businesses desiring secure and private financial transactions.

Monero, along with other privacy-focused cryptocurrencies like Zcash and Dash, represents an essential facet of the crypto ecosystem. They cater to a growing demand for digital currencies that offer more than just the decentralized ledger of blockchain but also a shield against the prying eyes of the public and institutions.

Balancing Privacy and Regulatory Compliance

However, the rise of privacy coins presents a complex challenge: balancing the legitimate need for privacy with regulatory compliance. Financial privacy is a legitimate concern for many, but it also raises questions about the potential misuse of these currencies for illicit activities. Regulatory bodies worldwide are grappling with this dilemma, trying to find a middle ground where the benefits of privacy coins can be harnessed without compromising legal and ethical standards.

Governments and regulatory agencies are exploring ways to ensure that the use of privacy coins adheres to financial regulations, including anti-money laundering (AML) and know your customer (KYC) policies. This evolving regulatory landscape is a tightrope walk, trying to protect individual privacy while preventing financial crimes. The future approach to mitigating their illicit use may pivot from focusing on the transactions themselves to addressing the real-life consequences and activities surrounding these transactions. As regulatory bodies and law enforcement agencies grapple with the inherent privacy of these cryptocurrencies, the likelihood increases that the solution lies not in penetrating the veil of transactional anonymity, but rather in leveraging evidentiary facts derived from outside the financial sphere.

This approach would involve focusing on the tangible outcomes and physical aspects of illicit activities, using traditional investigative methods and intelligence gathering. This shift recognizes the limitations inherent in monitoring privacy-centric blockchain transactions and instead places emphasizes on the broader context of the transactions, tracking illegal activities through non-financial clues, patterns, and real-world implications. This method acknowledges the robust privacy protections of these coins, while also upholding legal and ethical standards in combatting the unlawful actions linked to their use.

Decentralized Applications (DApps)

In the realm of cryptocurrency and blockchain technology, Ethereum has emerged as a pivotal platform, not just as a digital currency but as the backbone of a vast ecosystem of decentralized applications (DApps) and smart contracts. Its role extends far beyond that of a mere transactional currency, venturing into the realms of decentralized finance, governance, and beyond.

Ethereum's groundbreaking contribution lies in its ability to execute smart contracts – self-executing contracts with the terms of the agreement directly written into lines of code. This functionality forms the core of Ethereum's offering, enabling a multitude of applications that go far beyond simple currency transactions. These smart contracts are immutable, transparent, and executed automatically upon meeting set conditions, thereby eliminating the need for intermediaries.

The platform has become a fertile ground for developers to build a diverse range of DApps, from gaming to decentralized finance (DeFi) platforms. These applications leverage Ethereum's smart contract capability to create complex financial instruments, execute automatic trading strategies, or even run entire autonomous organizations.

From Ownership to Autonomous Organizations

The use cases of Ethereum's technology are as diverse as they are revolutionary. In the realm of digital art and collectibles, Ethereum's blockchain has facilitated the rise of non-fungible tokens (NFTs), which are used to prove ownership and authenticity of unique digital assets. In the world of DeFi, Ethereum powers lending platforms, decentralized exchanges, and yield farming protocols, all operating without traditional financial intermediaries.

Moreover, Ethereum has paved the way for the creation of Decentralized Autonomous Organizations (DAOs). These entities operate entirely on blockchain, with smart contracts automating their governance and operational processes. DAOs represent a radical shift in organizational structure, promising a future of decentralized and democratized decision-making.

The Concept and Implications of CBDCs

In the future financial landscape, Central Bank Digital Currencies (CBDCs) represent a significant development, intertwining the traditional banking system with the innovative realm of digital currencies. These state-backed digital currencies bring forth a complex blend of opportunities and challenges.

CBDCs are digital forms of fiat money, issued and regulated by a nation's central bank. Unlike decentralized cryptocurrencies like Bitcoin or Ethereum, CBDCs are centralized and maintain all the characteristics of traditional fiat currency, merely in digital form. This evolution in currency form is driven by the need for modernization of the financial system, aiming to enhance the efficiency of transactions and broaden financial inclusion. However, the implications of CBDCs extend far beyond mere technical advancements, touching upon economic policies, privacy concerns, and financial control.

One of the most significant concerns surrounding CBDCs is the potential they hold for increased government control and surveillance. With transactions being digital and traceable, there is a possibility for governments to monitor and regulate financial activities closely. When intertwined with systems like a social credit score, CBDCs could become tools for controlling not just financial transactions but also influencing social behaviors. This level of oversight and control leads to concerns about privacy and personal freedom.

For some, the concept of CBDCs raises parallels with prophetic concerns such as those in Christian eschatology, particularly the concept of the 'Mark of the Beast,' which is often associated with control and surveillance in the end times. This concept, deeply rooted in Christian theology, is often interpreted as a symbol of ultimate control and dominance in the end times. Revelation 13:16-17 (KJV) states, *"And he causeth all, the small and the great, and the rich and the poor, and the free and the bond, that there be given them a mark on their right hand, or upon their forehead; and that no man should be able to buy or to sell, save he that hath the mark, even the name of the beast or the number of his name."* Such interpretations view the advent of CBDCs as a step towards a future where financial and social compliance could be enforced through technology.

Utopian Views on CBDCs

On the flip side, there are utopian perspectives on CBDCs, highlighting their potential to revolutionize the financial system. Proponents argue that CBDCs could make financial transactions more efficient, secure, and inclusive. They could reduce the costs and complexities of money circulation, streamline governmental disbursements like social welfare payments, and potentially stabilize national financial systems.

The Dystopian Views on CBDCs

Conversely, the dystopian view emphasizes the risks of centralization and loss of privacy. In this view, CBDCs could lead to unprecedented surveillance capabilities, with governments having access to detailed information about individual spending habits. The fear is that this could lead to a scenario where financial transactions are not just monitored but could also be censored or directly controlled, leading to a loss of financial freedom and autonomy.

The emergence of CBDCs is a multifaceted development in the world of finance, embodying the potential to reshape the way monetary systems function. While they promise enhanced efficiency and modernization, they also bring forth significant debates about privacy, control, and the balance of power between individuals and the state. The future of CBDCs will likely be defined by how these competing perspectives and concerns are navigated and addressed.

Those Without Digital Access

In the future landscape of financial assets, the relationship between traditional value stores like gold and silver and emerging digital currencies presents an intriguing dynamic, reflecting both historical continuity and potential inversion.

Gold and silver have been cornerstones of wealth and monetary systems for millennia, prized for their intrinsic value, scarcity, and durability. Cryptocurrencies, although a digital and recent invention, share some of these key attributes, especially in terms of scarcity and durability in a digital sense. However, cryptocurrencies offer additional features such as ease of transfer, divisibility, and in some cases, anonymity, which are not inherent in physical metals.

Digital currencies have the potential to revolutionize financial inclusion, particularly for the unbanked population. They offer easier access to financial services without the need for traditional banking infrastructure. However, this potential is tempered by the need for digital access and literacy. For those without internet access or digital devices, the benefits of cryptocurrencies remain out of reach.

The future might witness an interesting inversion in the realm of value stores. Historically, gold and silver were the currencies of affluence, while lacking such assets was a mark of the less affluent. In a future dominated by digital currencies and assets, this relationship could invert. As digital currencies become more mainstream, accessible, and regulated, they might become the standard for the majority, including those in higher economic strata.

In such a scenario, physical gold and silver coins could become the refuge of the disenfranchised or those wary of digital systems - a reversal of their historical role. This inversion would be akin to how horses, once the primary mode of transportation for all, have become a luxury for the wealthy, while cars, once a luxury, are now ubiquitous and essential for the broader population.

This potential shift could lead to a future where traditional and digital value stores coexist, each serving different segments of society based on access, preference, and trust. Gold and silver may emerge as tangible assets for those skeptical of digital systems or lacking digital access, while cryptocurrencies could serve as the primary financial tool for the digitally connected world.

In Summary

Imagine a future where financial transactions are not bound to a single dominant currency but are instead facilitated by a diverse array of digital currencies. In this future, Bitcoin continues to reign as a store of value and a means for significant transactions, while altcoins like Dogecoin and Ethereum serve daily commerce and smart contract applications, respectively. Privacy coins like Monero offer anonymity where needed, and CBDCs (Central Bank Digital Currencies) represent the digital evolution of national currencies. This rich tapestry of cryptocurrencies enables users to choose the best fit for their needs, creating a dynamic and flexible financial environment.

The advent of this multi-crypto ecosystem harbors the potential for greater inclusivity. Digital currencies can offer financial services to the unbanked and underbanked, breaking down barriers that traditional banking systems have maintained. By providing diverse options, this ecosystem caters to a broader range of economic activities and preferences, from global trade to personal finance, and from large institutional transactions to micro-payments.

In this envisioned future, the power of choice and agency takes center stage. Individuals and businesses can select currencies based on factors like stability, privacy, transaction speed, and purpose of use. This freedom of choice extends beyond mere transactional convenience; it reflects personal values, privacy preferences, and even philosophical inclinations toward decentralization and financial sovereignty.

The multi-crypto future is not without its challenges – regulatory hurdles, technological advancements, and market volatility all play a role in shaping this landscape. However, the overarching theme is one of empowerment and opportunity. As we progress into this future, the role of education and informed decision-making becomes paramount. Users must navigate this complex ecosystem with a clear understanding of the risks and benefits associated with each type of cryptocurrency.

The multi-crypto future promises a financial ecosystem that is more adaptable, inclusive, and reflective of individual needs and values. It's a vision of a world where financial transactions are not just a matter of economic necessity but also a choice that aligns with one's lifestyle, beliefs, and goals. As this future unfolds, it holds the promise of transforming not just how we transact, but also how we perceive the very nature of value and wealth.

Chapter 18
Visions of Potential Futures in Cryptonomics

In the future, where the digital and physical intertwine seamlessly, the advent of cryptocurrency has reshaped not just economies but the very fabric of daily life. This new era, buoyed by the principles of Cryptonomics, is where we find ourselves as we begin our narrative journey.

We are introduced to a family, our protagonists in this story, living in a world where cryptocurrency is not a speculative asset or a technological novelty but an integral part of everyday life. This family, diverse in its generational makeup, navigates this crypto-enabled world with a familiarity that comes from years of living within a system where digital currencies are as commonplace as the traditional money of the past.

Their daily life is a tapestry of interactions, both mundane and significant, all underpinned by the use of various cryptocurrencies. From the morning purchase of coffee paid for in Dogecoin to the smart contract-enabled home automation system running on Ethereum, their day is a showcase of how deeply integrated these technologies have become in personal and societal spheres.

This narrative is set against a socio-economic backdrop that has evolved significantly from our current day. The global economy in this future is a decentralized yet interconnected network of transactions and exchanges, where the barriers of traditional financial systems have given way to a more inclusive and efficient model.

Their story is not just about the technology they use but about how these tools shape their interactions, their opportunities, and their challenges. It's a glimpse into a future that is not too distant, where the full potential of cryptocurrency and blockchain technology is realized, changing not just how we buy and sell but how we live, work, and interact with each other in a world where the lines between the digital and physical are increasingly blurred.

The Thompsons

As dawn breaks over the city, the first rays of sunlight filter through the windows of the Thompson family's smart home, a cozy yet modern dwelling nestled in the heart of a bustling metropolis. The Thompsons, a family of four, are early adopters of technology, living in a world where digital currencies and decentralized organizations are the norm.

The family begins their day with a routine that might seem familiar yet is subtly interwoven with advanced technology. Over breakfast, they discuss the agenda for their local Decentralized Autonomous Organization (DAO) meeting. This particular DAO, a digital platform for community governance, allows residents like the Thompsons to participate directly in decision-making processes that impact their neighborhood.

In this world, traditional town hall meetings have transformed into digital forums where community members engage, debate, and make decisions using blockchain technology. The Thompsons, active participants in their local DAO, are preparing to vote on a new community project. The proposal? A sustainable urban garden is to be built on a vacant lot nearby.

As they finish their breakfast, the family logs into the DAO platform. The interface is user-friendly, displaying the proposal details and a simple mechanism for casting votes. Each family member, including the two teenage children, has a say. They review the project plan, discussing its merits and implications for the community. The conversation is lively, with the younger Thompsons bringing fresh perspectives on environmental sustainability and communal spaces.

After some debate, they cast their votes, each contributing to the decision-making process in real time. The blockchain records their input transparently and immutably, ensuring that every voice is heard and counted. This direct and transparent method of governance empowers the Thompsons and their neighbors, fostering a sense of ownership and responsibility towards their community.

As the day progresses, they receive a notification on the DAO platform: the urban garden project has been approved by a majority vote. The family feels a sense of accomplishment and connection to their neighbors, even those they haven't met in person. The DAO has not only streamlined the decision-making process but has also strengthened the community's bond by encouraging active participation and transparent communication.

In this future, the Thompsons are more than just residents; they are active contributors to the shaping of their community's future. DAOs have revolutionized governance, making it more inclusive, efficient, and democratic. For the Thompson family, this is just another day in a life where technology and community intertwine, creating a harmonious balance between individual agency and collective action.

As the day unfolds, the Thompsons turn their attention to an aspect of life that's as crucial as it is complex – managing their finances. However, in their world, this task is streamlined and efficient, thanks to the advent of Decentralized Finance, or DeFi.

Gone are the days of traditional banking, where a visit to the local branch and mountains of paperwork were the norms. Now, sitting in the comfort of their living room, with a few taps on their holographic interface, the Thompsons access a suite of DeFi platforms that handle everything from daily transactions to long-term investments.

Mr. Thompson, a graphic designer, reviews the family's digital asset portfolio. He shuffles through various DeFi platforms, each offering a range of services, from high-yield savings accounts to diverse investment opportunities. He reallocates some assets into a new blockchain-backed real estate venture, excited by the potential growth and the ease with which he can diversify their investments.

Meanwhile, Mrs. Thompson, a freelance software developer, explores a peer-to-peer lending platform. She's considering a loan to upgrade her home office. Traditional credit checks and loan approvals are things of the past. Here, smart contracts automate the process, assessing her creditworthiness using transparent criteria and disbursing funds almost instantly.

The most enthusiastic about DeFi, however, are the Thompson children, Mia and Alex. They're not just passive observers of their parents' financial activities; they're actively learning and participating. Mia, the older sibling, has been saving her allowance in a DeFi savings account, watching it grow steadily thanks to the higher interest rates these platforms offer. Alex, meanwhile, has been dabbling in cryptocurrency trading, using a small fund allocated by his parents as an educational tool.

This evening, the Thompsons gather for a family finance session, a weekly ritual where they discuss and learn about managing money in the DeFi world. Today's lesson is about risk management and the importance of diversification. The children listen intently as their parents explain complex concepts in simple terms, using interactive tools and simulations provided by their DeFi platforms.

For the Thompsons, DeFi isn't just a tool for financial management; it's an educational platform that empowers each family member with knowledge and independence. They have bid farewell to the constraints of traditional banking, embracing a world where financial control is in their hands, guided by the principles of transparency, inclusivity, and autonomy. In this digital age, they are not just passive consumers of financial services but active participants in a financial ecosystem that is constantly evolving and adapting to their needs.

In the Thompson household, the world of Non-Fungible Tokens, or NFTs, is an integral part of daily life. These unique digital assets have reshaped their understanding and interaction with both the physical and digital worlds, blending the realms of ownership, art, and technology.

Mr. Thompson recently acquired a vintage digital watch, its value rooted not in the physical item but in the NFT linked to it. This token verifies its authenticity and his ownership, revolutionizing the way he collects and cherishes items. Similarly, Mrs. Thompson, a software developer, uses NFTs to secure and prove ownership of her digital creations, appreciating the indisputable proof of ownership and the ease of transferring rights when needed.

For the Thompson children, NFTs are woven into both play and education. Mia, working on a school art project, plans to mint her creation as an NFT. But she goes a step further, linking it to a copyright contract, showcasing not only her artistic talent but also her savvy understanding of digital rights management. This project is more than an assignment; it's a foray into the world of digital asset management and intellectual property.

Alex, the younger sibling, immerses himself in the world of gaming, where in-game items are NFTs. He trades and sells these tokens, learning about digital economics in a practical, engaging manner.

The family's digital display wall is a testament to their journey in this NFT-enriched world. It showcases a rotation of digital art and tokenized family photographs, each piece narrating a unique story. Their collection is a blend of art and technology, memories preserved not just in pixels but in blockchain entries.

As they gather around, reminiscing about the stories behind each piece, the Thompsons reflect on how NFTs have reshaped their experiences. Mia's school project, now a minted NFT with its copyright contract, sits proudly among the collection, symbolizing a new era where art, technology, and legal savvy coalesce.

In the Thompson family's life, cryptocurrencies have become the norm for transactions of all kinds. Their usage of various digital currencies is tailored to the nature of their transactions, reflecting the versatility and diversity of the crypto world.

For significant purchases and savings, the family relies on Bitcoin. Its stability and wide acceptance make it their go-to choice for large-scale transactions. When Mr. Thompson bought their electric car, he made the payment in Bitcoin. It was a smooth, efficient process, free from the hassles of traditional financing. Similarly, their long-term savings are held in Bitcoin, its digital gold status giving them a sense of security and investment growth over time.

For everyday purchases, Dogecoin and other similar cryptocurrencies are the Thompsons' favorites. The family appreciates the low transaction fees and the currency's suitability for quick, routine transactions. Whether it's buying groceries or paying for Mia's online music lessons, Dogecoin offers them a convenient and efficient means of payment. These altcoins, designed for daily use, have become as common as cash once was, making digital transactions a seamless part of their day-to-day life.

Privacy is a value the Thompsons hold dear, especially in transactions that require discretion. For these, Monero becomes their currency of choice. Its privacy-centric design ensures that their financial data remains confidential, providing peace of mind for sensitive transactions. When Mrs. Thompson made an anonymous donation to a whistleblower protection charity, she used Monero, knowing her privacy would be protected.

This diverse cryptocurrency ecosystem has given the Thompson family financial freedom and flexibility like never before. Bitcoin, Dogecoin, Monero, and others each play their unique role, together creating a comprehensive digital financial toolkit. This approach to transacting has not just simplified their financial dealings but has also instilled in them a nuanced understanding of the strengths and applications of different cryptocurrencies.

In their world, the choice of currency is not just a matter of value; it's a reflection of the purpose, privacy, and priorities of each transaction. This nuanced approach to using various cryptocurrencies is a testament to the adaptability and ingenuity of the Thompsons, symbolic of how families in the future might navigate the complex yet rewarding world of digital finance.

In the Thompson household, the application of smart contracts has streamlined many aspects of their daily routine, particularly in managing their home and finances. Ethereum's blockchain, renowned for its robust smart contract capabilities, plays a central role in this automation.

Automating recurring payments and subscriptions is one of the important areas where the Thompsons have leveraged the power of smart contracts. Utility bills, streaming services, and even Mia and Alex's weekly allowances are managed through automated Ethereum-based contracts. Once set up, these contracts autonomously execute transactions when due, ensuring payments are never missed and removing the need for manual intervention. This automation not only saves time but also provides peace of mind, knowing that all regular commitments are taken care of efficiently and reliably.

In the realm of real estate and contractual agreements, the Thompsons have also embraced the advantages of smart contracts. Their rental agreement, for example, is a digital contract on the Ethereum blockchain. It details terms and conditions transparently and executes actions like rent collection and maintenance requests automatically. When they recently had to renew their lease, the process was as simple as validating a transaction on the blockchain, a stark contrast to the cumbersome paperwork of the past.

Similarly, service contracts for household needs such as cleaning, landscaping, or repairs are managed via smart contracts. These agreements detail service expectations, schedules, and payments, all encoded on the blockchain. This setup not only streamlines the management of household services but also fosters a trust-based relationship with service providers. Payments are released upon the satisfactory completion of services, as verified by both parties, ensuring fairness and transparency.

Even their home automation system is integrated with smart contracts. The system autonomously adjusts heating, lighting, and security settings based on predefined conditions and preferences. When the family leaves the house, a smart contract ensures that all lights are off, the security system is activated, and the thermostat is adjusted to an energy-saving mode. This integration of smart contracts into home automation not only enhances the family's comfort and security but also optimizes energy usage, reflecting their commitment to sustainability.

For the Thompsons, smart contracts are not just a technological innovation; they are an integral part of their daily lives, bringing efficiency, transparency, and trust to various household operations. This adoption of Ethereum-based smart contracts illustrates how blockchain technology can transcend the financial sector, permeating into practical, everyday applications, making life simpler and more secure for families like the Thompsons.

The Wei Family

In a parallel narrative, far from the Thompsons' experience, lies the story of the Wei family in a future version of China, where life is deeply intertwined with a government-controlled digital currency system tied to a comprehensive social scoring system.

The Weis live in a bustling metropolis under a regime where every transaction and social interaction feeds into their social score. This number dictates their access to services, employment opportunities, and even social status. This score is influenced by their behavior, financial transactions, and compliance with government policies, all monitored through the Central Bank Digital Currency (CBDC).

Every morning, Mr. Wei checks the family's social score, knowing that a dip below certain thresholds could mean restrictions on travel, shopping, or his children's education opportunities. He recalls an instance when a late payment on a utility bill due to a misunderstanding had resulted in a temporary score drop, leading to a tense few days where all non-essential purchases were automatically declined until the matter was rectified.

Their financial transactions are entirely in the state-controlled CBDC, leaving a transparent trail for the government to monitor. Purchasing certain items, or even shopping at specific stores, can affect their score. Mrs. Wei often finds herself double-checking if buying a particular book or viewing a foreign film online might impact their score negatively.

Their daughter, Li, a high school student, is acutely aware of the impact of social scores. Her dream of attending a top university is as contingent on her academic performance as it is on maintaining a high family score. She's cautious about her social media posts and the discussions she engages in, aware that dissenting opinions could jeopardize her future.

This system also influences the Wei family's social interactions. They are cautious about forming new friendships, as associations with individuals having low social scores could bring down their own. Which could put a cap on how much they can save or lower their savings interest rate. The total control of the state-run CBDC has led to a society where trust and open communication are often compromised, replaced by a constant, underlying concern about maintaining one's social standing.

In their world, privacy is a concept of the past. Surveillance cameras with facial recognition are omnipresent, and every digital action is logged and analyzed. The CBDC system, while efficient and modern, is also a tool for control and compliance, making financial freedom a notion that the Wei family can only dream of.

Their life is a cautionary tale of a society where the intersection of finance and governance crosses into the realm of personal freedom and autonomy. It's a stark contrast to the Thompsons' world, where digital currencies empower and liberate. For the Weis, it represents a reality where technology becomes an apparatus of control that is deeply embedded in the very fabric of their daily existence.

Chapter 19
Embracing the Revolution

As we draw to the close of this exploration into the burgeoning world of cryptocurrencies and their profound impact on our economic systems, it's crucial to reflect on the journey we've embarked upon. This journey has taken us through the intricate evolution of money, from its earliest forms to the digital currencies that are now beginning to reshape our financial landscape.

The evolution of money and economic theories has been a tale of constant adaptation and transformation. Throughout history, money has evolved to meet the changing needs of societies and economies, from barter systems to gold coins, from paper notes to digital transactions. This evolution is not just a story of technological advancement but also one of shifting economic theories and practices. It is a narrative that demonstrates humanity's relentless pursuit of more efficient, secure, and equitable financial systems.

In recent years, the emergence of Bitcoin and blockchain technology has inaugurated a new chapter in this evolution. These innovations have challenged traditional notions of what money is and can be. Bitcoin, with its decentralized nature, has introduced a radical shift away from centralized financial systems, offering a glimpse into a future where financial autonomy and peer-to-peer transactions are the norms. Blockchain technology, the backbone of Bitcoin, has demonstrated that it's possible to have a secure, transparent, and immutable ledger, a feature that could revolutionize not just currency but all forms of digital transactions.

The role of cryptocurrency in modern finance is multi-faceted and continually evolving. It has opened up new avenues for investment, transactions, and even the way we conceptualize value and trust in the financial sector. Cryptocurrencies have shown potential in addressing some of the longstanding challenges of the current financial system, including issues related to accessibility, efficiency, and security. They have also sparked important conversations around the control and governance of monetary systems, prompting us to reconsider who holds power over our financial infrastructures and decisions.

As we step into the future, the lessons learned from the history of money, combined with the revolutionary potential of Bitcoin and blockchain, pave the way for a reimagined financial landscape. This journey encourages us to continually reassess and adapt our economic theories and practices in light of emerging technologies and their profound implications for our financial futures.

In our exploration of the transformative landscape of modern finance, it becomes increasingly clear that the economic theories of the past, while foundational, must be revisited and revised in the context of today's digital age. The advent of cryptocurrencies and the broader concept of cryptonomics compel us to reconsider and reshape our economic understanding.

Historically, economic systems have been shaped and reshaped in response to various challenges, whether it be the collapse of currency systems, the Great Depression, or the more recent global financial crisis of 2008. These events have consistently highlighted the vulnerabilities in existing economic structures and theories, emphasizing the need for adaptability and resilience. Learning from these historical economic challenges is crucial. They serve as reminders of the importance of developing economic theories that are not only robust but also flexible enough to adapt to changing circumstances and technologies.

As we find ourselves in the throes of the digital age, the integration of cryptocurrencies and blockchain technology into our economic systems represents a seismic shift, akin to the industrial or digital revolutions of the past. This new era of cryptonomics challenges traditional concepts of currency, value, and financial intermediation. It calls for an economic paradigm that embraces decentralization, digitalization, and the democratization of financial systems. Adapting to this new reality requires not only technological proficiency but also a fundamental rethinking of economic principles and models.

The future role of decentralized finance (DeFi) is particularly significant in this regard. DeFi extends the ethos of cryptocurrencies into broader financial services, offering a glimpse into a world where financial services are not governed by central authorities but are instead built on transparent, decentralized, and equitable systems. This shift holds the potential to redefine banking, lending, investment, and even insurance, making them more accessible, efficient, and secure. DeFi challenges the centralized financial power structures and opens up a realm of possibilities for innovation and inclusion.

As we look forward, it is evident that our journey through the realms of Bitcoin, blockchain, and cryptonomics is not just about understanding new technologies. It is, more importantly, about recognizing their potential to reshape our economic theories and practices. It is about preparing ourselves for a future where finance is not just a tool for economic growth but also a means for achieving greater equity, empowerment, and economic democracy.

In this era of rapid technological advancement and economic transformation, the importance of deepening our understanding of Bitcoin and its underlying technology cannot be overstated. Bitcoin, more than just a new type of currency, represents a paradigm shift in our approach to financial systems. It challenges conventional thinking about money, prompting questions about what constitutes value, trust, and decentralization in finance. For those curious about this evolving landscape, delving deeper into Bitcoin's mechanics, history, and potential applications is the first step towards full comprehension and informed participation.

The crypto revolution, heralded by the advent of Bitcoin and other digital currencies, opens up a plethora of opportunities that extend far beyond mere financial transactions. It is a movement that encompasses various aspects of society and technology, including blockchain applications in sectors like healthcare, supply chain management, and digital identity verification. For entrepreneurs, developers, investors, and everyday users, this revolution presents an array of avenues to explore – from developing new blockchain-based solutions to investing in cryptocurrencies, or simply using them as a medium of exchange.

The dynamic nature of this field, with its ongoing developments and evolving regulatory landscapes, underscores the importance of continual learning and engagement. Staying updated with the latest trends, understanding the risks involved, and actively participating in the discourse surrounding cryptocurrencies are essential for anyone keen to be part of this revolution. This engagement is not just about personal or financial gain; it is also about contributing to the shaping of a more inclusive, transparent, and efficient financial future.

As we venture into this new era, let us approach it with an open mind and a spirit of inquiry. Whether one is a skeptic or an enthusiast, the crypto revolution, much like the internet revolution before it, is poised to bring about significant changes. Engaging with it, understanding it, and contributing to its responsible growth can ensure that its benefits are maximized and its risks, mitigated. In doing so, we not only embrace the revolution brought about by Bitcoin and its counterparts but also play a part in steering its course toward a future that aligns with our collective ideals of equity, efficiency, and economic empowerment.

The transformative potential of cryptocurrencies, spearheaded by Bitcoin, is vast and multi-dimensional, extending its reach far beyond the realm of traditional finance. This transformative impact is not just theoretical but is being felt and observed in real-time across various sectors and global financial systems.

Bitcoin's influence on global finance has been nothing short of revolutionary. As the first and most prominent cryptocurrency, it has challenged the traditional understanding of what a currency can be. It operates independently of central banks, offering a decentralized alternative to state-issued fiat currencies. This has not only introduced a new asset class for investors but has also opened doors for financial inclusion, allowing people without access to traditional banking to participate in the global economy. Moreover, Bitcoin's underlying technology, blockchain, has spurred a rethinking of how financial transactions are recorded and verified, promising increased efficiency and transparency.

However, with these groundbreaking prospects come challenges, especially in the realm of decentralized finance (DeFi). DeFi aims to recreate and improve upon traditional financial services like lending, borrowing, and investing, but without the intermediaries such as banks. While it offers unprecedented accessibility and potential for finantial democratization, it also faces hurdles like regulatory uncertainties, scalability issues, and concerns over security and user protection. Addressing these challenges is crucial for DeFi to realize its full potential and to ensure a stable and secure financial environment for all participants.

The influence of cryptocurrency is also rapidly expanding across various other sectors. In supply chain management, blockchain technology is being used to enhance transparency and traceability. In the energy sector, it facilitates peer-to-peer energy trading and sustainable practices. Even in areas like healthcare and art, the impact of digital currencies and blockchain technology is becoming increasingly evident, whether it's through the secure sharing of medical records or the use of Non-Fungible Tokens (NFTs) to authenticate and sell digital art.

The transformative potential of cryptocurrencies and blockchain technology extends well beyond digital transactions. They are reshaping the very fabric of global finance and numerous other industries, ushering in a new era of technological advancement, economic empowerment, and cross-sector innovation. As this transformation unfolds, it presents a unique opportunity for individuals, businesses, and governments to engage with, adapt to, and ultimately shape a new financial and technological landscape.

In the ever-evolving landscape of cryptocurrencies, the power of individual involvement cannot be overstated. Each participant in the cryptocurrency ecosystem, be it a seasoned investor, a tech enthusiast, or a casual user, plays a vital role in shaping the future of these digital assets. This involvement is not merely about personal gain or technological curiosity; it's about actively contributing to the development of a financial system that values autonomy, decentralization, and democratization.

The role of individuals in shaping the future of cryptocurrencies is multifaceted. It involves participating in the market, engaging in community discussions, contributing to technological advancements, and even simply using cryptocurrencies for everyday transactions. Each action, whether big or small, contributes to the larger narrative of what cryptocurrencies can become and how they can function in our society. By actively participating, individuals can influence the direction of cryptocurrency development, advocating for features and regulations that align with the community's needs and values.

Advocating for financial autonomy and decentralization is another critical aspect of individual involvement. Cryptocurrencies, at their core, are about reducing reliance on centralized financial institutions and intermediaries. They empower users to have direct control over their financial transactions and assets. By embracing and advocating for these principles, individuals help promote a financial ecosystem that is more inclusive, efficient, and resistant to censorship and control by a few dominant players. This advocacy is crucial in ensuring that cryptocurrencies remain tools for financial empowerment rather than becoming co-opted by existing financial structures.

Resisting centralized control is an essential component of the cryptocurrency ethos. While centralization can bring efficiency and order, it often comes at the cost of freedom, privacy, and equality. Cryptocurrencies offer a way to push back against this trend, providing a means for secure, private, and independent financial transactions. By supporting cryptocurrencies, individuals stand up against the concentration of power in the hands of a few large institutions and governments, advocating for a more balanced and equitable financial landscape.

The role of individuals in the world of cryptocurrencies is both powerful and necessary. It is about more than just adopting a new technology; it's about participating in a movement that seeks to redefine the very nature of financial autonomy and freedom. By getting involved, individuals can help ensure that the future of cryptocurrencies is shaped by a diverse range of voices and interests, ultimately leading to a more fair and decentralized financial system for all.

As we stand on the brink of a new era shaped by the rise of cryptocurrencies, it's vital to envision the future they are likely to create. The impact of these digital currencies on future economies is poised to be profound and far-reaching. Cryptocurrencies, with their decentralized nature, offer a radical departure from traditional financial systems. They promise to democratize finance, making it more accessible and equitable. This could lead to a significant reduction in transaction costs, increased speed and efficiency in payments, and greater financial inclusion worldwide, especially for those who are currently unbanked.

Blockchain technology, the backbone of cryptocurrencies, is set to play a critical role in building transparent and fair systems. Its ability to maintain an immutable, distributed ledger of transactions ensures transparency and accountability, qualities often lacking in current financial systems. This could revolutionize not just finance but also other sectors like supply chain management, healthcare, and voting systems, ensuring that data is handled in a transparent and tamper-proof manner. Blockchain's potential to bring about greater transparency and fairness in various systems is immense, but it hinges on thoughtful implementation and widespread adoption.

However, as we embrace these technological advances, ethical considerations and the need for balanced technological advancement cannot be overlooked. The rise of cryptocurrencies and blockchain presents questions regarding privacy, security, and regulatory oversight. There is a delicate balance to be struck between harnessing the benefits of these technologies and ensuring they do not exacerbate existing inequalities or create new forms of digital divide. The decentralized nature of cryptocurrencies can also lead to challenges in governance and accountability, making it crucial to develop ethical frameworks and guidelines for their use.

Moreover, the environmental impact of cryptocurrency mining, particularly for Bitcoin, has raised concerns and calls for sustainable practices. This presents an opportunity for innovation in developing more energy-efficient consensus mechanisms and leveraging renewable energy sources.

In envisioning the future, it is essential to consider these various dimensions and ensure that the advancements in cryptocurrencies and blockchain technology are harnessed responsibly and ethically. This future should be one where technology serves to enhance economic systems, making them more inclusive, efficient, and fair, while also addressing the critical ethical and environmental challenges posed by these new technologies. As we move forward, it is the collective responsibility of developers, users, regulators, and all stakeholders to guide this technological evolution in a direction that benefits all members of society.

In the transformative journey of cryptocurrencies, it is not just the technology itself but the active participation of individuals that drives real change. This final chapter serves as a call to action for everyone, from seasoned tech enthusiasts to newcomers curious about this space, to become a part of the cryptocurrency movement. Your involvement, no matter the scale, contributes to shaping a future where finance is more equitable, accessible, and under the control of the many rather than the few.

Participating in the cryptocurrency movement does not necessarily mean investing your savings into digital currencies. It can be as simple as educating yourself and others about the potential and risks of this technology, engaging in community discussions, or using cryptocurrencies for small, everyday transactions. For those with the technical know-how, it could mean contributing to the development of blockchain technologies or creating user-friendly applications that make cryptocurrencies more accessible. Every action helps in building a more robust and diverse ecosystem.

Understanding and embracing your role in a fair financial system is crucial. Cryptocurrencies offer an opportunity to rethink our relationship with money and the financial institutions that govern it. By choosing to use cryptocurrencies, advocating for sensible regulations, or even just by engaging in discussions about their potential and limitations, you are taking a step towards a system where financial control and power are distributed more widely rather than concentrated in the hands of a few.

Finally, this journey is not one to be undertaken in isolation. The strength of cryptocurrencies and blockchain technology lies in their inherent design to be decentralized and community-driven. This is a call to emphasize community and collaboration, to participate in a way that not only benefits you as an individual but also contributes to the collective good. The cryptocurrency movement, at its core, is about building a more inclusive and equitable financial system, and this can only be achieved through collective effort and collaboration.

As we move forward into this new era of digital finance, let us do so with a sense of purpose, responsibility, and community. The revolution of cryptocurrencies is not just a financial or technological one; it is a movement towards a future where each of us has a role to play in shaping a fair and just financial system. This is your invitation to be a part of this exciting and transformative journey.

As we conclude our journey through the intricate and evolving world of cryptocurrencies and their impact on our economic systems, its essential to recognize that we stand on the cusp of a new economic era. This era is marked by rapid technological advancements and a shift towards decentralization in finance. Embracing this change requires an openness to new ideas, a willingness to adapt, and a commitment to understanding how these technologies can reshape not just our financial transactions, but also our broader societal structures.

In this new era, the principles underpinning cryptocurrencies and blockchain technology – transparency, decentralization, and empowerment – offer a blueprint for a future that values financial inclusivity and individual autonomy. It is a future where the traditional barriers to financial access are dismantled, and where each individual has greater control over their financial destiny. As we embrace this change, we must also remain mindful of the challenges it presents, including regulatory uncertainties, ethical considerations, and the need for sustainable development within this space.

Reflecting on the journey laid out in these pages, my final thoughts return to the core message of this book: the coming economic revolution. This revolution is not just about the rise of Bitcoin or the technical intricacies of blockchain. It is about a fundamental rethinking of how we view and interact with money. It is about challenging long-held assumptions and being open to new possibilities in our financial systems.

"The Coming Economic Revolution" is more than a forecast; it is a call to action. It is an invitation to participate in shaping this new era, to contribute your voice to the ongoing discussions, and to play a part in steering the future of finance. As we move forward, let us do so with both caution and optimism, recognizing the potential of these technologies to create a more equitable and empowered society.

In closing, whether you are a skeptic or an enthusiast, a technologist or a layperson, the world of cryptocurrencies and blockchain presents an opportunity to engage with one of the most significant economic developments of our time. As we embark on this journey together, let us remain open-minded, inquisitive, and collaborative, driving forward the coming economic revolution.

Appendix

Appendix A
Resources for Further Learning

The journey into understanding Bitcoin and economic theory doesn't end with this book. For those eager to explore deeper, the following resources provide a wealth of information, offering various perspectives and insights into the world of economics and cryptocurrency.

Books

"The Big Three in Economics: Adam Smith, Karl Marx, and John Maynard Keynes"
by Mark Skousen

"The Big Three in Economics" is a compelling and insightful exploration into the lives and ideas of three of the most influential economists in history: Adam Smith, Karl Marx, and John Maynard Keynes. Authored by Mark Skousen, a renowned economist and writer, this book delves into the foundational economic theories that have shaped our modern world.

"The Wealth of Nations" by Adam Smith

Regarded as the foundational text of classical economics, Smith's work provides insight into the nature and causes of the wealth of nations, discussing concepts like division of labor, productivity, and free markets.

"A Tract on Monetary Reform"
by John Maynard Keynes
A significant work in the field of economics, this book offers Keynes's perspectives on monetary policy, inflation, and the international monetary system, which are essential to understanding modern economic theories.

"The Bitcoin White Paper"
by Satoshi Nakamoto
The foundational document for understanding Bitcoin. This white paper, titled "Bitcoin: A Peer-to-Peer Electronic Cash System," lays out the conceptual and technical foundations of Bitcoin.

Websites

CoinDesk (coindesk.com)
Provides news, analysis, and information on Bitcoin and other digital currencies, making it a valuable resource for staying updated on the latest in the cryptocurrency world.

Investopedia (investopedia.com)
Offers comprehensive explanations of economic concepts, market analysis, and reviews of financial products. It's particularly useful for understanding the basics of economics and finance.

Look Into Bitcoin (lookintobitcoin.com)
They provide basic on-chain data for free and a complete set of tools for Bitcoin analysis for a fee.

Bitcoin.org (bitcoin.org)
A primary resource for information directly related to Bitcoin, offering detailed insights into how Bitcoin works, its features, and its potential impact on the global economy.

YouTube Channels

Coinbureau (youtube.com/@CoinBureau)
The Coin Bureau is your go-to informational portal to the cryptocurrency galaxy. Our mission is to facilitate the mass adoption of cryptocurrency through education, one person at a time.

Digital Asset News (youtube.com/@DigitalAssetNews)
The top cryptocurrency and digital asset news

Podcasts

"The Pomp Podcast" by Anthony Pompliano
Focused on business, finance, and Bitcoin, this podcast features interviews with figures in the cryptocurrency world and discussions on economic trends.

"Planet Money" by NPR
Makes complex economic stories interesting and understandable. Though not exclusively focused on Bitcoin or cryptocurrencies, it offers valuable insights into economic principles and current events.

"Unchained" by Laura Shin
A podcast that dives into the world of cryptocurrencies and blockchain, exploring how these technologies are changing the digital landscape.

Forums

BitcoinTalk (bitcointalk.org)

The largest and one of the oldest message boards dedicated to blockchain and cryptocurrencies. Here, enthusiasts and experts discuss various aspects of Bitcoin and other digital currencies.

r/Bitcoin on Reddit (reddit.com/r/bitcoin)

A vibrant community where members share news, ask questions, and discuss the technical and economic aspects of Bitcoin.

CryptoCompare (cryptocompare.com)

A community-driven platform where users can discuss and analyze cryptocurrency trading and investment.

These resources offer a starting point for your journey. Each book, website, podcast, and forum provides a unique perspective, contributing to a well-rounded understanding of Bitcoin, blockchain technology, and economic theory. As you explore these resources, remember that the fields of economics and cryptocurrency are ever-evolving, and staying informed is key to deepening your understanding.

Appendix B
A Note on the Use of AI and this Book.

In the creation of this book, AI (Artificial intelligence) has played a significant role akin to that of a co-author, bringing its unique strengths to the writing process while also presenting particular challenges that needed careful navigation. The use of AI in writing assistance has been a dance of balancing its capabilities with its limitations. On the one hand, AI's ability to generate content quickly has been invaluable. However, this has come with the need to address certain inherent shortcomings, such as its tendency to have a limited attention span and be occasionally repetitive in its content. Additionally, managing the AI's stylistic flair to align with the overall tone and voice of the book requires a discerning human touch.

In general, AI's role can extend beyond mere assistance; it can also act as a catalyst, particularly for aspiring authors. For those facing initial hesitations about writing or struggling with where to start, AI can provide a jumping-off point, generating ideas and outlines that can be expanded upon and personalized. It can be incredibly empowering for writers with limited experience, offering a framework upon which they can build their narratives and express their ideas.

However, the utilization of AI in literary creation brings with it a set of ethical considerations. Paramount among these is the need to guard against the mass production of low-quality content. With AI's ability to generate volumes of text, there is an ever-present risk of flooding the literary market with content lacking depth or meaningful insight. Therefore, responsible use of AI is crucial. This means ensuring that AI is used to enhance and supplement human creativity, not replace it. AI's role is to support the author's vision, providing a foundation upon which the author's ideas could be constructed rather than dominating the creative process.

In essence, the AI used in the creation of this book was a tool, a digital ally, that helped bridge gaps in the creative process and eased the journey from conception to completion. Yet, it was the human touch that remained at the helm, guiding the narrative, shaping the ideas, and imbuing the content with authenticity and purpose.

In the journey of crafting this book, the collaboration between human creativity and artificial intelligence was both intricate and nuanced. The overall concept and scope of the book were initially defined by me, the author, laying the foundational vision and objectives that would guide the entire project. This initial vision was crucial as it set the parameters within which AI tools were utilized, ensuring that the technology served the book's goals and not the other way around.

The AI's role was used in generating the initial book outline. This outline, while a product of AI's algorithms, was not accepted blindly. It underwent a rigorous process of curation and editing. I meticulously reviewed each section and concept, expanding and refining them to ensure they aligned with the overarching themes and objectives of the book. There were instances where the AI suggested new information that the author was initially unaware of. These were rare, but when they occurred, I took care to verify their relevance and accuracy before including them in the final outline. This process ensured that the book remained true to its intended course, reflecting a thoughtful and purposeful exploration of the subject matter.

Furthermore, the AI was employed in drafting initial outlines for various paragraphs and sections. These AI-generated drafts provided a starting point, a raw material that I shaped and molded. Each paragraph's outline was carefully reviewed, edited, and often rewritten to ensure it met the standards of clarity, coherence, and engagement that I had set for the book. This iterative process of refinement was crucial in maintaining the quality and consistency of the narrative.

Lastly, the act of integrating AI-generated content was accompanied by a verification process. Every piece of content produced by the AI was scrutinized for accuracy and originality. I employed plagiarism checks to ensure the uniqueness of the material and edited or rewrote sections to uphold the integrity and authenticity of the book. This rigorous process was necessary to ensure that the book not only provided accurate and valuable information but also maintained a unique voice and perspective.

It is imperative to underscore that the essence, the voice, and the vision of the work are intrinsically mine, the author's. Every aspect of this book, from its inception to its final form, has been a manifestation of my intent and understanding. The decision to use AI in the creation process was driven by the desire to enhance and augment this vision, not to replace or overshadow it.

The presentation of the material within these pages reflects my deliberate choices. The organization of topics, the depth of exploration, and the tone of the narrative were all carefully crafted to offer readers a clear, coherent, and engaging journey. This careful curation was essential in ensuring that the book remained true to my objectives as its creator.

Recognizing the importance of clarity in conveying ideas, especially in subjects as intricate as cryptocurrency and economics, I employed repetition strategically. It served as a tool to reinforce key concepts, aiding in the reader's comprehension and retention of the material. This repetition was used judiciously, ensuring it served the purpose of education and kept the overall flow and readability of the text intact.

In essence, AI was a valuable tool in this creative endeavor, a digital assistant that helped polish and refine my ideas. It aided in the structuring of content, provided initial drafts, and suggested perspectives that might have otherwise been overlooked. However, I made the final decisions on content, structure, and presentation, ensuring that the book remained a true reflection of my knowledge and insights.

As you, the reader, reach the end of this journey, I hope that the book has not only enriched your understanding of cryptocurrency and its implications for the future but also sparked your imagination. The world of crypto is vast and ever-evolving, and this book will offer you a valuable perspective on its potential to shape our global economic landscape.

OpenAI. (2023). ChatGPT [Large language model]. https://chat.openai.com

Appendix C
About the Author

Donald D. Dienst (pronounced Deen st), at the age of 55, stands as a testament to the power of self-directed learning and the pursuit of diverse interests. With a rich and varied career that spans several disciplines, he embodies the spirit of a Renaissance man in the digital age.

His early fascination with programming led him to explore the intricacies of coding, laying the foundation for his future endeavors in the tech industry. Over the years, his passion for understanding how things work drove him to master the fields of electronic engineering and product design. In these areas, he applied his self-acquired knowledge to develop innovative solutions and products, showcasing his ability to blend technical skill with creative vision.

But Donald's talents are not confined to engineering and design alone. He also ventured into the realm of game design, where he combined his technical expertise with a flair for creativity and storytelling. This foray into game design highlights his versatile skill set and his ability to engage and entertain through interactive experiences.

Throughout his career, he has exemplified a lifelong commitment to learning and adapting. His journey is a powerful reminder of how passion and self-motivation can drive success across multiple fields. His diverse expertise not only reflects his personal interests but also gives him a unique perspective on the convergence of technology, design, and entertainment.

Appendix D
Other DienstNet Products

Quester's Keep - A Competitive RPG Board Game

Quester's Keep is a board game for 1 to 6 players. A game of exploration, character development, engaging combat, survival, and suspense! Where everyone is in a race to complete their secret quests and survive long enough to get their treasures back to the keep.

Available at www.Amazon.com

The Pico C API Functionary

The ultimate function reference guide for Pico C programmers. This comprehensive book serves as a dictionary of the functions in the Pico C API, providing you with an invaluable resource to navigate and utilize the extensive functionality of the language. Inside you'll find a well-organized collection of functions, presented in a clear and concise manner.

Available at www.Amazon.com

Raspberry Pi Pico W Documentation Compilation

A compilation of the Raspberry Pi Pico W Pinout PDF, the Raspberry Pi Pico W Product Brief PDF, and the Raspberry Pi Pico W Datasheet PDF. For people that enjoy using and working with printed manuals. The information in this compilation is available for free in PDF format directly from Raspberry Pi. 33 full color pages.

Available at www.Lulu.com/shop

www.ingramcontent.com/pod-product-compliance
Lightning Source LLC
Chambersburg PA
CBHW071336210326
41597CB00015B/1473